The
Deep Ecology
Movement

The Deep Ecology Movement

An Introductory Anthology

Edited by
**Alan Drengson
& Yuichi Inoue**

North Atlantic Books
Berkeley, California

Published by
North Atlantic Books
Berkeley, California

This is issue #50 in the *Io* series.

Cover and book design by Paula Morrison
Printed in the United States of America by Allura Printing, Inc.

The Deep Ecology Movement: An Introductory Anthology is sponsored and published by the Society for the Study of Native Arts and Sciences (dba North Atlantic Books), an educational nonprofit based in Berkeley, California, that collaborates with partners to develop cross-cultural perspectives, nurture holistic views of art, science, the humanities, and healing, and seed personal and global transformation by publishing work on the relationship of body, spirit, and nature.

This book was partially funded by the Foundation for Deep Ecology.

North Atlantic Books' publications are available through most bookstores. For further information, visit our website at www.northatlanticbooks.com or call 800-733-3000.

ISBN-13: 978-1-55643-198-2

Library of Congress Cataloging-in-Publication Data
Drengson. Alan R.
    The deep ecology movement : an introductory anthology / Alan
Drengson, Yuichi Inoue.
        p. cm.
    Includes bibliographical references.
    ISBN 1-55643-198-8
    I. Deep ecology.   I. Inoue, Yuichi.   II.Title.
GE195.D74   1995
363.7—dc20   95-3367
                                                                              CIP

From the Elders

"It is the story of all life that is holy and is good to tell, and of us two-leggeds sharing in it with the four-leggeds and the wings of the air and all green things; for these are children of one mother and their father is one spirit."

—Black Elk, Sioux Elder

"There are mountains hidden in jewels, there are mountains hidden in marshes, mountains hidden in the sky; there are mountains hidden in mountains. There is a study of mountains hidden in hiddenness."

—Dogen, Zen Master

"Behold the lilies of the field, how they grow; neither do they toil, nor do they spin: And yet I say unto you, that even Solomon in all his glory was not arrayed as one of these."

—Jesus of Nazareth

"All things . . . have their own will and their own way and their own purpose; this is what is to be respected."

—Rolling Thunder, Healer

"A thing is right when it tends to preserve the integrity, stability, and beauty of the biotic community. It is wrong when it tends otherwise."

—Aldo Leopold, Ecologist

This anthology is dedicated to all who seek to dwell wisely within their place respecting the community of all beings upon whom they depend. May all beings flourish!

Table of Contents

Section II. Elaboration

Section III. Major Topics

Acknowledgements

The editors are grateful for the support of the Foundation for Deep Ecology, whose assistance made this anthology possible. Thanks to Eileen King for proof-reading, and to Sandra Leland and Brenda Bulmer at Genie Computing, for word processing services. Also thanks to Richard Grossinger, Marianne Dresser, and the other people at North Atlantic Books for their assistance in expediting and producing this book. We thank our families for their patience with the time this work took us away from them. Special thanks to Bill Devall for his encouragement and advice, to Doug Tompkins for his support, and to Danny Moses for his efforts on behalf of this book. We thank all of the contributors for making their essays available for publication. We also thank the following publishers for permission to use the articles cited in the following entries:

Arne Naess' "The Shallow and the Deep, Long-Range Ecology Movement: A Summary," *Inquiry* 16 (1973), 95–100 (Reprinted by permission of Scandinavian University Press, Oslo, Norway);

Arne Naess' "The Apron Diagram" synthesized by Prof. Inoue with permission of Prof. Naess. It was drawn from "Deep Ecology and Ultimate Premises," *The Ecologist* 18 (1988), 4/5, 128–131; and "Intuition, Intrinsic Values and Deep Ecology," *The Ecologist* 14 (1984), 5/6, 201–203 (Permission by *The Ecologist*, Agriculture House, Bath Road, Sturminster Newton, Dorset DT10 IDU, England);

Arne Naess' "Self-Realization: An Ecological Approach to Being in the World," from *The Trumpeter* 4 (1987), 3, 35–42, and from Murdoch University where it was given as an address;

Arne Naess' "Ecosophy T" from his *Ecology, Community and Lifestyle: Outline of an Ecosophy*, translated and revised by David Rothenberg, Cambridge University Press, Cambridge, 1989, 196–210;

Arne Naess' and George Sessions' "Platform Principles of the

Acknowledgements

Deep Ecology Movement," from Bill Devall and George Sessions, *Deep Ecology: Living as if Nature Mattered*, Gibbs Smith, Salt Lake City, © 1985, 69–73;

George Sessions' "Arne Naess and the Union of Theory and Practice," from *The Trumpeter* 9 (1992), 2, 73–76;

Gary Snyder's "Re-inhabitation" was published in *The Old Ways*, City Lights Books, San Francisco, 1977, 57–66, permission from author (This article is included in Snyder's volume: *A Place in Space* [Washington, D.C., Counterpoint Press, 1995]);

Alan Drengson's "Shifting Paradigms: From the Technocrat to the Planetary Person," from *Environmental Ethics* 3 (1980), 221–240;

Bill Devall's "The Ecological Self" from his *Simple in Means, Rich in Ends: Practicing Deep Ecology*, Gibbs Smith, Salt Lake City, © 1988, 38–72;

Freya Mathews, "Conservation and Self-Realization: A Deep Ecology Perspective," from *Environmental Ethics* 10 (1988), 347–55;

Warwick Fox's "Transpersonal Ecology and the Varieties of Identification," from his *Toward a Transpersonal Ecology: Developing New Foundations for Environmentalism*, Shambhala, Boston, 1990, 249–268 (From *Toward a Transpersonal Ecology* by Warwick Fox. © 1990 by Warwick Fox. Reprinted by arrangement with Shambhala Publications, Inc., 300 Massachusetts Avenue, Boston, MA 02115);

David Rothenberg's "A Platform of Deep Ecology" from *The Environmentalist* 7 (1987), 185–190;

Michael Zimmerman's "Feminism, Deep Ecology, and Environmental Ethics" from *Environmental Ethics* 9 (1987), 21–44;

Patsy Hallen's "Making Peace with Nature: Why Ecology Needs Feminism" from *The Trumpeter* 4 (1987), 3, 3–14;

Dolores LaChapelle's "Ritual is Essential" from Bill Devall and George Sessions, *Deep Ecology: Living as if Nature Mattered*, Gibbs Smith, Salt Lake City, 1985, 247–250, originally published in *In Context* 5 (1984), rights held by the author;

Pat Fleming's and Joanna Macy's "The Council of All Beings" from *Thinking Like a Mountain: Toward a Council of All Beings*, John Seed et al., Editors, New Society Publishers, Philadelphia, 1988,

79–90 (Permission by New Society Publishers, 4527 Springfield Avenue, Philadelphia, Pennsylvania, 19143, ph. 1-800-333-9093);

Gary Snyder's "Ecology, Place, and the Awakening of Compassion" from *Turning Wheel*, Spring 1994, 12–15, was also given as a talk at a Village Council of All Beings in Leh, Ladakh (This article is included in Snyder's forthcoming volume: *A Place in Space* [Washington, D.C., Counterpoint Press, 1995]);

John Rodman's "Four Forms of Ecological Consciousness Reconsidered" from *Ethics and the Environment*, D. Scherer and T. Attig, Editors, Prentice-Hall, Englewood Cliffs, N.J., 1983, 82–92;

Andrew McLaughlin's "For a Radical Ecocentrism" from his *Regarding Nature: Industrialism and Deep Ecology*, SUNY Press, Albany, N.Y., 1993, 197–205, 210–212, 217–225 (Reprinted from *Regarding Nature: Industrialism and Deep Ecology* by Andrew McLaughlin, by permission of the State University of New York Press, Copyright © 1993 State University of New York);

The Ecoforestry Institutes of the U.S.A. and Canada for permission to reprint their materials including the text of *The New York Times* advertisement.

About the Authors

(in order of appearance)

Arne Naess is Emeritus Professor of Philosophy at the University of Oslo in Norway. His office is at the Center for Environment and Development at the University. He is the author of *Ecology, Community and Lifestyle*.

George Sessions is a philosophy instructor at Sierra College in Rocklin, California. He is the co-author of *Deep Ecology*.

Gary Snyder is a poet and essayist as well as a faculty member at the University of California in Davis. He is the author of *Turtle Island* and *The Practice of the Wild*.

Alan Drengson (see the note under Editors).

Bill Devall is a professor of Sociology at Humboldt State University in Arcata, California. He is the co-author of *Deep Ecology*, and the author of *Simple in Means, Rich in Ends*.

Freya Mathews is a professor of philosophy at LeTrobe University in Australia. She is the author of *The Ecological Self*.

Warwick Fox is a Research Fellow in Environmental Studies at the University of Tasmania in Australia. He is the author of *Toward a Transpersonal Ecology*.

David Rothenberg is Assistant Professor of Philosophy at the New Jersey Institute of Technology in Newark. He is the author of *Hand's End*.

Michael Zimmerman is a professor of philosophy at Newcomb College at Tulane University in New Orleans. He is the author of *Heidegger's Confrontation with Modernity*.

Patsy Hallen is a professor of philosophy at Murdoch University in Australia. She is the author of numerous essays.

Dolores LaChapelle is Director of the Way of the Mountain Learning Center in Silverton, Colorado. She is the author of *Sacred Land, Sacred Sex*.

Pat Fleming is a facilitator and leads Councils of All Beings. She is the founder of Earth Care College in England. She is one of the editors of *Thinking Like a Mountain*.

Joanna Macy is Adjunct Professor at California Institute for Integral Studies in San Francisco, and is a workshop leader and conducts Councils of All Beings. She is the author of *World as Lover, World as Self*. She is an editor of *Thinking Like a Mountain*.

John Rodman is a professor of political science at Pitzer College in Claremont, California. He is the author of numerous works.

Andrew McLaughlin is a professor of philosophy at Lehman College in New York. He is the author of *Regarding Nature*.

The Ecoforestry Institute is a nonprofit, charitable, educational institution organized in the U.S. and Canada committed to teaching ecologically responsible forest use. See their addresses at the end of the Introduction to this book.

About the Editors

Alan Drengson, is a philosopher at the University of Victoria, in Victoria, B.C., Canada. He is the author of *Beyond Environmental Crisis* and *The Practice of Technology*, and Senior Editor of *The Trumpeter*.

Yuichi Inoue is a professor of environmental studies at Nara Sangyo University in Nara, Japan. He is the translator of the Japanese version of Brian Tokar's *The Green Alternative*.

Preface

For several years we have both independently thought of putting together an anthology of readings on the Deep Ecology Movement. Such an anthology has been very much needed in environmental studies and philosophy courses, but also as a resource for a wider audience.

When Prof. Inoue came to the University of Victoria he suggested to Prof. Drengson that they work on an introductory anthology of deep ecology literature which could be translated into Japanese. The collection slowly evolved and became a dual language project. The first edition of this collection will appear in English. It will later appear in Japanese.

We are happy to offer readers this sustained introduction to issues and problems of great substance and urgency. The philosophical and other analyses presented herein are not dull reading. The topics discussed are of fundamental human importance.

Arne Naess has remarked that one feature of the deep ecology movement is that its supporters ask deeper questions. We trust that the essays in this collection will stimulate readers to do their own deep questioning and begin to explore their own ecological Selves, if they have not already been doing so. If they have made these discoveries on their own, let these essays stimulate their further journeys and explorations.

But above all, we hope that this brief survey of some of the literature of the deep ecology movement will lead to support for the integrity of the natural order so that all beings may flourish. In this we hope that reflections stimulated by this book will lead to practical actions and changes in lifestyle compatible with ecocentric values and respect for all life.

An important note to readers: The authors of the essays included in this book have asked us to inform readers that in many cases they would not articulate their views today in the way presented by these papers. In some instances these essays were written more than thirty years ago. The authors' views have evolved, times have changed. We wanted nonetheless to offer to readers a sense of the historical development of recent work in support of the deep ecology movement. The bibliography at the end of this book will aid readers on their further explorations. Items are marked to indicate different levels of immersion in the literature.

Introduction

Alan Drengson and Yuichi Inoue

1. Formation

In 1973, Norwegian philosopher Arne Naess published the summary of a lecture delivered at the third World Future Research Conference (Bucharest, September 1972). In it, he articulated the distinction between the shallow and the deep approaches to environmentalism. He characterized the latter by seven major points. This summary, the opening article of this anthology, is regarded as the "original" paper on the grassroots movement to ecocentrism. It introduced the term *Deep Ecology Movement* and presented a basic characterization of it. Thus, the words "deep ecology" have been identified with this movement since the early 1970s. The first Earth Day was commemorated in 1970. The United Nations Conference on Human Environment was held in Stockholm in 1972, and the first "Oil Crisis" came in 1973.

2. Arne Naess

Arne Naess is a highly respected and distinguished person. He is a prominent scholar, internationally recognized for his voluminous, innovative, and insightful work on general and specific questions in philosophy. He was appointed to the Chair of Philosophy at the University of Oslo at the age of twenty-seven in 1939, and retained that position until 1969, when he took early retirement. Naess has led academic work in philosophy and social research in Norway for a long time. After his resignation of the professorship, Naess put specific emphasis on developing professional work on ecosophy (ecological philosophy and wisdom). One of his major publications on ecosophy is *Ecology, Community and Lifestyle* (translated and edited with the help of David Rothenberg).

Naess is well known for his dedication to nonviolent civil resistance. He has practiced Gandhian nonviolence, under such critical conditions as the Norwegian resistance to Nazi occupation, as well as in ecological protest actions against large-scale development projects. Naess is a distinguished mountaineer, known for leading the first ascent of the highest peak (7692 meters) in the Hindu Kush, among other climbs. He is an ardent hiker and skier with a deep love and sense of wonder for nature. Finally, it should be noted that Naess has fascinated many people with his warm-hearted friendliness and humor, multifaceted interests and knowledge, ardent love of and devotion to nature and truth, and his fairness and open-mindedness. George Sessions' paper in this anthology describes with a personal touch major characteristics of Naess' philosophy and lifestyle.

3. Introduction to North America

Despite the 1973 publication of Naess' original paper, the term "deep ecology" was barely referred to in North America until the 1980s. Warwick Fox divides the development of the deep ecology movement into three stages: the latency period (1973 to 1980); the honeymoon period, when virtually all responses were positive (1980 to 1983/84); and the mature period, when the deep ecology movement began to attract critical as well as positive commentaries, during which the framework for the movement has been substantially deepened and enlarged (1983/84 to today).

George Sessions and Bill Devall were two of the first people to become aware of the importance of Naess' original paper. They started referring to and elaborating on deep ecology ideas in their own writings in 1977. They thereby helped to introduce the concepts of the deep ecology movement to North America. *Deep Ecology*, their joint work published in 1985, was the first major book on the deep ecology movement by writers other than Naess. It played a significant role in popularizing the ideas of the movement. The two most important journals in which writings on the deep ecology movement appear are: *Environmental Ethics* (Department of Philosophy, PO Box 13496, University of North Texas, Denton, TX 76203-3496, USA) which started in 1979, and *The Trumpeter: Journal of Ecosophy* (LightStar: PO Box 5853, Stn. B, Victoria, BC V8R 6S8, Canada) which is devoted to the deep ecology movement and began publishing in 1983.

4. What is Deep Ecology?

Naess' distinction between the shallow and the deep approaches, made more than twenty years ago, has become even more relevant today. In the decade of the 1990s, environmental issues have become widely seen as one of the greatest challenges facing humankind, and no one, not even the expansionist (i.e., economic-growth oriented), can ignore this. This fact was highlighted by the United Nations Conference on Environment and Development (UNCED or the Earth Summit) in Rio de Janeiro in June of 1992. So-called corporate ("shallow") environmentalism still dominates the mainstream. It advocates continuous economic growth and environmental protection by means of technological innovation (such as catalytic converters), "scientific" resource management (such as sustained yield forestry), and mild changes in lifestyle (such as recycling). It avoids serious fundamental questions about our values and worldviews; it does not examine our sociocultural institutions and our personal lifestyles. This mainstream technological approach has to be clearly distinguished from the deep ecology approach, which in contrast examines the *roots* of our environmental/social problems. The deep approach aims to achieve a fundamental ecological transformation of our sociocultural systems, collective actions, and lifestyles.

Unfortunately, the two approaches, shallow and deep, are too often confused in the public mind, and this confusion adversely affects efforts to achieve an environmentally sustainable, socially equitable, and spiritually rich way of life. It is true that supporters of the deep ecology movement and those of the shallow technological approach share many of the same concerns, and they should cooperate where necessary and appropriate. This is well understood by environmentalists and so the term "reform" is often used as an alternative to "shallow." Such confusion, however, can divert attention from what needs to be done. Therefore, it is important to understand how supporters of the deep ecology movement have elaborated, developed and applied the elements of their views in relation to Naess' original distinction. This helps us to develop clarity so we can concentrate our best efforts on practical solutions.

Gary Snyder's paper on reinhabitation is a seminal essay like Naess' summary paper. There are strong lines of connection between the two. Snyder's articulations helped to inspire the back to the land movement in North America. The reinhabitation paper published here was seminal to bioregionalism (an important element in the ecology movement that originated in North America) and it implicitly conveys the shallow-deep continuum for discovering the inherent values of nature. Snyder's writings crystalized the Eastern influence on the Pacific side of North America. We can detect this coastal-montane voice in other writers in this anthology, such as Alan Drengson and Bill Devall.

Alan Drengson's article in this anthology, published in 1980, is an early, original exposition of Naess' distinction. In it, Drengson contrasts the person-planetary and technocratic paradigms, in terms of the deep-shallow typology. His paper was written in 1978 developing the distinctions between the technocratic and pernetarian (*persons* in *networks* of plane*tary* relationships) paradigms. When the editor of *Environmental Ethics*, Eugene Hargrove, sent him a copy of Naess' summary paper in 1979, Drengson realized that the shallow-deep description fit his own discovery of the same value continuum but was verbalized in simpler terms. He incorporated this into his paper. The paper received wide distribution and helped to bring a number of different approaches and fields together in interdisciplinary studies of the environmental crisis as a deep cultural one. In 1983 Drengson founded *The Trumpeter: Journal of Ecosophy* specifically to disseminate ecosophies in support of the deep ecology movement.

In 1984, Naess and Sessions jointly developed a preliminary platform or a set of basic principles, which was offered as a minimum description of the general features of the deep ecology movement. We can understand this platform (and the characterization made in

Naess' 1973 paper) as the core tenets of the movement. These describe the general, broad meaning of "deep ecology" as a world-wide, grassroots movement. However, the term "deep ecology" has also been used in a more narrow sense, that is, to refer to Naess' own "Self-realization" thesis, a particular approach (or ecosophy) within eco-centric, deep environmentalism. This approach to ecosophy has been the narrow meaning of "deep ecology." Because of this dual use, the term "deep ecology" has seemed confusing to those unfamiliar with this historical background. Naess explained early that many ecosophies, or ultimate premises, could support the platform principles, and the platform principles could yield many sorts of practical applications (see Chapter 2).

Naess has been careful in his own use of these terms. He uses the term "deep ecology movement" to refer to a broad ecocentric grassroots effort, as contrasted with an anthropocentric, technocratic approach, to achieve an ecologically balanced future. He sees this effort as a social and political movement. On the other hand, he reserves the term "deep ecology" (but more often uses "Ecosophy T") for referring to his own specific ecological philosophy of ultimate premises centered on "Self-realization" achieved through wider identification with one's ecological context. *Here, "deep ecology" is used to refer to a philosophy, not a movement.* The distinction between these two phrases is important, but in the environmental literature people often use the term "deep ecology" to refer to what Naess and other supporters call "the deep ecology movement." People also use "deep ecology" to refer to "Ecosophy T," "the identification thesis," and "the Self-realization thesis," that is, Naess' own ecological philosophy. (Warwick Fox calls Naess' own ecosophy a form of transpersonal ecology.) To avoid ambiguity, it would be better if we used the whole phrase "deep ecology movement" when referring to the movement and the other terms (not "deep ecology") when referring to specific ecosophies.

5. The "Apron" Diagram

The "Apron" Diagram included in this anthology is highly useful for understanding the relationship between Naess' own "deep ecology" philosophy and "the deep ecology movement." It was developed by Naess to represent a *total view* of the deep ecology movement. It is composed of four levels: 1. ultimate premises (philosophies/religions), 2. platform principles, 3. general views, and 4. practical/concrete decisions, presented in order of logical derivation. Naess advocates unity at the level of platform principles (Level 2), allowing plurality/diversity at the other levels. If we accept his formulation, we can say that the deep ecology movement involves efforts on four different levels, but it also can be represented by the above-mentioned "platform" principles alone, which belong to Level 2. On the other hand, the Self-realization thesis (Naess' own specific ultimate philosophy) belongs to the level of ultimate premises, Level 1. Thus, the Self-realization thesis is primarily a *philosophical* undertaking, whereas support of the platform principles is a political and social undertaking.

The Apron Diagram helps to clarify the whole spectrum of activity within the deep ecology movement, ranging from philosophical inquiries to concrete judgments and daily actions. It helps us to understand the relationship of a particular effort to the total framework of the movement. This understanding, then, lets us concentrate on any particular level without thereby losing the larger context and solidarity of the movement.

6. Self-Realization and Ecosophy T

The "Self-realization" thesis distinguishes Naess' philosophy from other ecocentric philosophical (mostly axiological) approaches. Naess advocates a psychological rather than a moralistic approach to environmentalism. His approach is to work on our inclinations, rather than preaching the subordination of our personal interest to an environmental ethic. He thinks that this can be done through expanding self beyond the boundaries of the narrow ego through the process of caring identification with larger entities such as forests, bioregions, and the planet as a whole. Naess' 1987 paper in this anthology explains in plain language what is meant by "self" with a small s (ego-self), "Self" with a capital S (ecological Self), and also what is meant by "extension of identification" to larger entities. His paper gives us a basic understanding of his own philosophy as an ecological way of being and his use of Gandhian philosophy as a source of personal inspiration, especially for non-violence in word and deed.

The norm of "Self-realization" and extension of identification is widely shared by many *explicit* supporters of the deep ecology movement, several of whom appear in this collection. Again, however, it must be stressed that it is not necessary to accept the Self-realization thesis to be a supporter of the deep ecology movement, because the thesis belongs to Level 1 (ultimate premises) but yields support for the broader deep ecology movement. At Level 1 pluralism is presupposed. One can support the platform from Shinto, Buddhist, or Christian ultimate premises, for example.

The Self-realization thesis has been widely studied and extensively elaborated upon by some supporters of the deep ecology movement. Three contributions in *Section II* of this anthology give examples of these efforts. Bill Devall develops the concept of "ecological self," as contrasted with "minimal self," and explicitly links the deep ecology movement with bioregionalism through the thesis of wide identification. Freya Mathews constructs an analysis to solve the seeming incompatibility between identification with the "ecocosm" (identification with the widest possible whole—the cosmic self) and conservation of individual life-forms, local ecosystems, and biosphere, thereby introducing a deeper dimension to conservation by means of the identification thesis. Warwick Fox advances Naess' Self-realization thesis by articulating the strengths and weaknesses of three forms of wide identification processes: the personal, the ontological, and the cosmological. Fox calls all of these approaches *transpersonal ecology.*

Naess' own Ecosophy T is a philosophical system composed of norms and hypotheses, with "Self-realization" as the ultimate norm; it represents Naess' own philosophical basis for his support of the deep ecology movement. Naess reached these basic norms by means of deep questioning, and he derives other norms from the basic ones in conjunction with factual hypotheses. The overview of Ecosophy T in this anthology helps us to understand the norm of "Self-realization!" (here the exclamation point is used to signal that this is a moral imperative) in its relationship to other important norms (such as Diversity!, Complexity!, Symbiosis!, Local autonomy!, No exploitation!, and Self-determination!). It serves to integrate Naess' 1973 and 1987 papers into a unified vision.

The term "ecosophy" is Naess' favorite alternative to "ecological philosophy," made up of eco- (originally means "house" or "habitation") as in ecology, and -sophy ("wisdom") as in philosophy (philo-sophia, love of wisdom). "T" is taken from Tvergastein, the name of Naess' tiny mountain hut in Norway. Over years of dwelling at Tvergastein, he has developed a deep spiritual kinship with the hut and its surroundings. This kinship constitutes an important part of his own ecosophy. By calling his ecophilosophy "Ecosophy T," Naess emphasizes that it is only his own ultimate philosophy, and that there can be other versions developed by others, such as Ecosophy A, Ecosophy B, and so on, each of which could support the platform principles of the deep ecology movement.

7. Platform Principles

As mentioned above, in 1984, Naess and Sessions prepared a preliminary platform composed of eight major principles. They thought that it appropriately characterized the deep ecology movement in a 200-word formulation. A recent version of it is included in this anthology. As with any political platform its main object is to bring into focus principles of general agreement on the wide range of issues facing the movement. The platform helps to inspire a sense of solidarity among local efforts in different contexts, which could otherwise appear isolated in a sea of divergent and global views. It also helps a wider public to understand the aims of the movement and to encourage grassroots support and participation in related efforts. Therefore, the platform principles avoid academic language and jargon. The aim of the platform is to promote clarity and consensus about the core principles shared by supporters of the deep ecology movement *despite their different backgrounds, whether these are philosophical, religious, occupational, or cultural.*

Naess and Sessions do not claim that their particular version of the platform is fixed or absolute. They emphasize its *preliminary* nature, and also invite others to propose their own version of the platform. David Rothenberg's paper in this anthology was a response to this invitation. Rothenberg's platform is compatible with the Naess/Sessions version, but there are also some differences. He tried to make his version accessible to an even wider audience by presenting the basic points in simpler words with a more obvious structure. Readers can decide which best describes their sense of the movement.

8. Critiques of the Deep Ecology Movement

As deep ecology (both identified with Naess' philosophy and as a movement) became more widely known and frequently referred to, various critiques emerged on the issues of science, anthropocentrism (versus ecocentrism), androcentrism (or patriarchal values), population control, and human liberty. Some of these critiques represent an emotional reaction based on a misinterpretation of the main tenets of the deep ecology movement, and as a result fail to be intellectually productive. However, others have helped supporters of the movement to clarify their own position and further develop ecocentric paradigms.

One interesting critique, for example, comes from the radical left wing of the ecology movement. Simply put, their criticism is that the deep ecology supporters' critique of the dichotomy between humans and the nonhuman world—represented by the industrial paradigm— results in lack of sufficient attention to such critical issues as the domination of humans by other human beings and also the exploitation of women by men: both of these, it is claimed, are located at the very root of today's ecological (environmental and social) crisis. Hence, these critics say that the deep ecology movement should give explicit priority to deep questioning about the roles of capital, the nation-state system, and sociocultural institutions based on traditional patriarchal values, in creating the crisis. They say this must precede deep questioning about human relationships to nature. For example, unlimited population growth logically results in ecological disaster, but, they argue, it is inappropriate to talk about population control before addressing such critical social/political issues as the tremendous inequity in resource distribution between the poor and the rich, and between the nations of the South and those of the North, and women's right to control their reproduction. It needs to be stressed that these positions are not incompatible with the broad deep ecology movement. Moreover, many supporters of the movement have explicitly acknowledged the importance of these issues. They might differ with respect to priorities needed to achieve the *radical* ecological transformation required of human beings living in the late twentieth century, if global and local environmental disasters are to be avoided. These critiques contain many rich insights that supporters of the deep ecology movement have incorporated into their views and activities.

9. Various Concerns in the Deep Ecology Movement

Section III of this anthology contains discussions of some of the major themes of the deep ecology movement. Michael Zimmerman's paper responds to the main ecofeminist criticisms of the movement. He looks at the essentially complementary character of these two ecocentric approaches. Patsy Hallen gives a feminist critique of science and reveals the psychosexual roots of the current ecological crisis. She also emphasizes the complementarity of ecofeminism and the deep ecology movement.

Dolores LaChapelle explores the meaning of rituals as activities that focus on the relationship of human beings to the land as sacred. She observes that rituals are sophisticated methods for communication essential to a sustainable culture. Pat Fleming and Joanna Macy outline the structure of an experiential workshop/ritual, which helps people to identify with other life-forms and speak for them, so that they can express their awareness of the current ecological crisis, while deepening their motivation to act.

Gary Snyder represents the bioregional approach and the spiritual dimension of the deep ecology movement, as influenced by Zen and wilderness experience. His poetic essay invites us to explore the "old" ways, the ways of the spiritual traditions rooted in primal sacredness found through compassion in specific places.

John Rodman presents a useful typology for understanding contemporary environmentalism, which helps to clarify the details of the deep/reform continuum. He advocates an environmental ethic based on an ecological sensibility from which environmental and social issues are perceived as *essentially* interrelated.

Finally, Andrew McLaughlin emphasizes the necessity of achieving a fundamental unity between the social justice and the nature preservation traditions in radical environmentalism. Together they should challenge industrialism and prepare the ground for an ecocentric future. His paper concludes *Section III* by describing the future of ecocentric movements and the opportunities for radical cultural changes.

10. Future Prospects

As we have explained above, the deep ecology movement unites a variety of interests, concerns and a diversity of religions and philosophies. As the movement continues to develop, there will be more diversity in activism, theory, and culture. Both specialization (working on a specific matter) and convergence (linking individual efforts into a unitary vision) will continue to be required.

Recently, foundations supporting the deep ecology movement have been established for specific purposes. In North America, for example, the *Ecoforestry Institute* (PO Box 12543, Portland, OR 97212, USA; and PO Box 5783, Stn. B, Victoria, BC V8R 6S8, Canada) has been incorporated both in the United States and Canada on the platform principles of the deep ecology movement, so as to promote ecologically responsible forestry and forest use. The Institute put a full-page educational statement in *The New York Times*, in which it advocated departure from the conventional "industrial" forestry to a new ecocentric forestry that respects forests for their own intrinsic worth and uses them on a sustainable basis. The text of this advertisement is in the appendix of this anthology. The Ecoforestry Institute also publishes the *International Journal of Ecoforestry* (PO Box 5885, Stn. B., Victoria, BC V8R 6S8, Canada), which started in 1994 to focus on the practices, science and philosophies of ecologically responsible forest use. The *Foundation for Deep Ecology* (950 Lombard Street, San Francisco, CA 94133, USA) was established in 1990 to provide financial support for projects and activities in the deep ecology movement. The *Institute for Deep Ecology Education* (IDEE) (A Project of the TIDES Foundation: Box 2290, Boulder, CO 80306, USA) sponsors training programs and workshops, and does consulting on deep ecology curricula and education. The *Ecostery Foundation of North America* (A Project of TIDES): PO Box 5885, Stn. B, Victoria, BC V8R 6S8, Canada) was founded in 1989 to promote ecosophy and establish "ecosteries" [a new word formed by combining "eco-" from "ecology" and "-stery" from "monastery"] as places where ecological wisdom is learned, practiced, and taught, under rural, suburban, and urban settings. The *Wildlands Project* (PO Box 5365, Tucson, AZ 85703, USA) has put forth a comprehensive wildlands recovery proposal for preserving biological diversity and wilderness values. The project inspired a Wildlands Anthology project, the first book (called *Place of the Wild*) was published at the end of 1994 by Island Press. They also help to publish the journal *Wild Earth* (PO Box 455, Richmond, VT 05477, USA). Finally, there is the *Arne Naess Selected Works Project* (Harold Glasser, Coordinator ANSWP, Applied Science, University of California at Davis, Davis, CA 95616, USA). This project's aim is to publish Professor Naess' major works in up-to-date English translations. Several of the volumes will be of critical importance to ecophilosophy and the deep ecology movement. These foundations, institutions, and projects are just a few examples of the many promising developments of the deep ecology movement as part of a transition to diverse ecocentric cultures and lifestyles.

Section I

Arne Naess

The Shallow and the Deep, Long-Range Ecology Movement: A Summary

Arne Naess

Ecologically responsible policies are concerned only in part with pollution and resource depletion. There are deeper concerns which touch upon principles of diversity, complexity, autonomy, decentralization, symbiosis, egalitarianism, and classlessness.

The emergence of ecologists from their former relative obscurity marks a turning point in our scientific communities. But their message is twisted and misused. A shallow, but presently rather powerful movement, and a deep, but less influential movement, compete for our attention. I shall make an effort to characterize the two.

1. The Shallow Ecology movement:

Fight against pollution and resource depletion. Central objective: the health and affluence of people in the developed countries.

2. The Deep Ecology movement:

(1) Rejection of the human-in-environment image in favor of *the relational, total-field image*. Organisms as knots in the biospherical net or field of intrinsic relations. An intrinsic relation between two things A and B is such that the relation belongs to the definitions or basic constitutions of A and B, so that without the relation, A and B are no longer the same things. The total-field model dissolves not only the human-in-environment concept, but every compact thing-

in-milieu concept—except when talking at a superficial or preliminary level of communication.

(2) *Biospherical egalitarianism*—in principle. The 'in principle'
clause is inserted because any realistic praxis necessitates some killing,
exploitation, and suppression. The ecological field-worker acquires
a deep-seated respect, or even veneration, for ways and forms of life.
He reaches an understanding from within, a kind of understanding
that others reserve for fellow humans and for a narrow section of
ways and forms of life. To the ecological field-worker, *the equal right
to live and blossom* is an intuitively clear and obvious value axiom.
Its restriction to humans is an anthropocentrism with detrimental
effects upon the life quality of humans themselves. This quality
depends in part upon the deep pleasure and satisfaction we receive
from close partnership with other forms of life. The attempt to ignore
our dependence and to establish a master-slave role has contributed
to the alienation of humans from themselves.

Ecological egalitarianism implies the reinterpretation of the future-
research variable, 'level of crowding,' so that *general* mammalian
crowding and loss of life-equality is taken seriously, not only human
crowding. (Research on the high requirements of free space of certain mammals has, incidentally, suggested that theorists of human
urbanism have largely underestimated human life-space requirements. Behavioral crowding symptoms [neuroses, aggressiveness,
loss of traditions ...] are largely the same among mammals.)

(3) *Principles of diversity and of symbiosis.* Diversity enhances
the potentialities of survival, the chances of new modes of life, the
richness of forms. And the so-called struggle of life, and survival of
the fittest, should be interpreted in the sense of ability to coexist and
cooperate in complex relationships, rather than ability to kill, exploit,
and suppress. 'Live and let live' is a more powerful ecological principle than 'Either you or me.'

The latter tends to reduce the multiplicity of kinds of forms of
life, and also to create destruction within the communities of the
same species. Ecologically inspired attitudes therefore favor diversity of human ways of life, of cultures, of occupations, of economies.
They support the fight against economic and cultural, as much as

military invasion and domination, and they are opposed to the annihilation of seals and whales as much as to that of human tribes or cultures.

(4) *Anti-class posture.* Diversity of human ways of life is in part due to (intended or unintended) exploitation and suppression on the part of certain groups. The exploiter lives differently from the exploited, but both are adversely affected in their potentialities of self-realization. The principle of diversity does not cover differences due merely to certain attitudes or behaviors forcibly blocked or restrained. The principles of ecological egalitarianism and of symbiosis support the same anti-class posture. The ecological attitude favors the extension of all three principles to any group conflicts, including those of today between developing and developed nations. The three principles also favor extreme caution towards any overall plans for the future, except those consistent with wide and widening classless diversity.

(5) Fight against *pollution and resource depletion.* In this fight ecologists have found powerful supporters, but sometimes to the detriment of their total stand. This happens when attention is focused on pollution and resource depletion rather than on the other points, or when projects are implemented which reduce pollution but increase evils of the other kinds. Thus, if prices of life necessities increase because of the installation of anti-pollution devices, class differences increase too. An ethics of responsibility implies that ecologists do not serve the shallow, but the deep ecological movement. That is, not only point (5), but all seven points must be considered together.

Ecologists are irreplaceable informants in any society, whatever their political color. If well organized, they have the power to reject jobs in which they submit themselves to institutions or to planners with limited ecological perspectives. As it is now, ecologists sometimes serve masters who deliberately ignore the wider perspectives.

(6) *Complexity, not complication.* The theory of ecosystems contains an important distinction between what is complicated without any Gestalt or unifying principles—we may think of finding our way through a chaotic city—and what is complex. A multiplicity of more or less lawful, interacting factors may operate together to form a

unity, a system. We make a shoe or use a map or integrate a variety of activities into a workaday pattern. Organisms, ways of life, and interactions in the biosphere in general exhibit complexity of such an astoundingly high level as to color the general outlook of ecologists. Such complexity makes thinking in terms of vast systems inevitable. It also makes for a keen, steady perception of the profound *human ignorance* of biospherical relationships and therefore of the effect of disturbances. Applied to humans, the complexity-not-complication principle favors division of labor, *not fragmentation of labor*. It favors integrated actions in which the whole person is active, not mere reactions. It favors complex economies, an integrated variety of means of living. (Combinations of industrial and agricultural activity, of intellectual and manual work, of specialized and non-specialized occupations, of urban and non-urban activity, of work in city and recreation in nature with recreation in city and work in nature. . . .)

It favors soft technique and 'soft future-research,' less prognosis, more clarification of possibilities. More sensitivity towards continuity and live traditions, and—most importantly—towards our state of ignorance.

The implementation of ecologically responsible policies requires in this century an exponential growth of technical skill and invention—but in new directions, directions which today are not consistently and liberally supported by the research policy organs of our nation-states.

(7) *Local autonomy and decentralization.* The vulnerability of a form of life is roughly proportional to the weight of influences from afar, from outside the local region in which that form has obtained all ecological equilibrium. This lends support to our efforts to strengthen local self-government and material and mental self-sufficiency. But these efforts presuppose an impetus towards decentralization. Pollution problems, including those of thermal pollution and recirculation of materials, also lead us in this direction, because increased local autonomy, if we are able to keep other factors constant, reduces energy consumption. (Compare an approximately self-sufficient locality with one requiring the importation of foodstuff,

materials for house construction, fuel and skilled labor from other continents. The former may use only five percent of the energy used by the latter.) Local autonomy is strengthened by a reduction in the number of links in the hierarchical chains of decision. (For example, a chain consisting of local board, municipal council, highest subnational decision-maker, a state-wide institution in a state federation, a federal national government institution, a coalition of nations, and of institutions, e.g., E.E.C. top levels, and a global institution, can be reduced to one made up of local board, nation-wide institution, and global institution.) Even if a decision follows majority rules at each step, many local interests may be dropped along the line, if it is too long.

Summing up, then, it should, first of all, be borne in mind that the norms and tendencies of the Deep Ecology Movement are not derived from ecology by logic or induction. Ecological knowledge and the life-style of the ecological field-worker have *suggested, inspired, and fortified* the perspectives of the Deep Ecology Movement. Many of the formulations in the above seven-point survey are rather vague generalizations, only tenable if made more precise in certain directions. But all over the world the inspiration from ecology has shown remarkable convergencies. The survey does not pretend to be more than one of the possible condensed codifications of these convergencies.

Secondly, it should be fully appreciated that the significant tenets of the Deep Ecology Movement are clearly and forcefully *normative*. They express a value priority system only in part based on results (or lack of results, cf. point [6]) of scientific research. Today, ecologists try to influence policy-making bodies largely through threats, through predictions concerning pollutants and resource depletion, knowing that policy-makers accept at least certain minimum *norms* concerning health and just distribution. But it is clear that there is a vast number of people in all countries, and even a considerable number of people in power, who accept as valid the wider norms and values characteristic of the Deep Ecology Movement. There are political potentials in this movement which should not be overlooked and which have little to do with pollution and resource depletion. In plotting possible futures, the norms should be freely used and elaborated.

Thirdly, insofar as ecology movements deserve our attention, they are *ecophilosophical* rather than ecological. Ecology is a *limited* science which makes *use* of scientific methods. Philosophy is the most general forum of debate on fundamentals, descriptive as well as prescriptive, and political philosophy is one of its subsections. By an *ecosophy* I mean a philosophy of ecological harmony or equilibrium. A philosophy as a kind of *sofia* [or] wisdom, is openly normative, it contains *both* norms, rules, postulates, value priority announcements *and* hypotheses concerning the state of affairs in our universe. Wisdom is policy wisdom, prescription, not only scientific description and prediction.

The details of an ecosophy will show many variations due to significant differences concerning not only 'facts' of pollution, resources, population, etc., but also value priorities. Today, however, the seven points listed provide one unified framework for ecosophical systems.

In general system theory, systems are mostly conceived in terms of causally or functionally interacting or interrelated items. An ecosophy, however, is more like a system of the kind constructed by Aristotle or Spinoza. It is expressed verbally as a set of sentences with a variety of functions, descriptive and prescriptive. The basic relation is that between subsets of premises and subsets of conclusions, that is, the relation of derivability. The relevant notions of derivability may be classed according to rigor, with logical and mathematical deductions topping the list, but also according to how much is implicitly taken for granted. An exposition of an ecosophy must necessarily be only moderately precise considering the vast scope of relevant ecological and normative (social, political, ethical) material. At the moment, ecosophy might profitably use models of systems, rough approximations of global systematizations. It is the global character, not preciseness in detail, which distinguishes an ecosophy. It articulates and integrates the efforts of an ideal ecological team, a team comprising not only scientists from an extreme variety of disciplines, but also students of politics and active policy-makers.

Under the name of *ecologism*, various deviations from the deep movement have been championed—primarily with a one-sided stress on pollution and resource depletion, but also with a neglect of the great differences between under- and over-developed countries in favor of a vague global approach. The global approach is essential, but regional differences must largely determine policies in the coming years.

References

Commoner, B., *The Closing Circle: Nature, Man, and Technology*, Alfred A. Knopf, New York 1971.

Ehrlich, P.R. and A.H., *Population, Resources, Environment: Issues in Human Ecology*, 2nd ed., W.H. Freeman & Co., San Francisco 1972.

Ellul, J., *The Technological Society*, English ed., Alfred A. Knopf, New York 1964.

Glacken, C.J., *Traces on the Rhodian Shore: Nature and Culture in Western Thought*, University of California Press, Berkeley 1967.

Kato, H., "The Effects of Crowding" Quality of Life Conference, Oberhausen, April 1972.

McHarg, I.L., *Design with Nature*, 1969. Paperback 1971, Doubleday & Co., New York.

Meynaud, J., *Technocracy*, English ed., Free Press of Glencoe, Chicago 1969.

Mishan, E.J., *Technology and Growth: The Price We Pay*, Frederick A. Praeger, New York 1970.

Odum, E.P., *Fundamentals of Ecology*, 3rd ed., W.E. Saunders Co., Philadelphia 1971.

Shepard, P., *Man in the Landscape*, Alfred A. Knopf, New York 1967.

The Apron

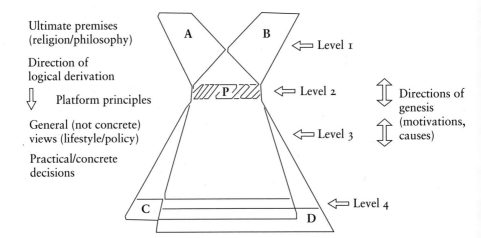

Ultimate premises
(religion/philosophy)

Direction of
logical derivation

Platform principles

General (not concrete)
views (lifestyle/policy)

Practical/concrete
decisions

A

B

⇐ Level 1

P

⇐ Level 2

⇐ Level 3

⇐ Level 4

C

D

Directions of
genesis
(motivations,
causes)

2

The Apron Diagram

Arne Naess

I see the Deep Ecology Movement as a total view comprising many levels in close contact with each other. To illustrate this I use a diagram. The "Apron Diagram" illustrates logical, as distinct from genetic, relations between views. By "logical relations," I mean verbally articulated relations between the premises and the conclusions. They move down the diagram in stages: some conclusions become premises for new conclusions. By "genetic relations," I refer to influences, motivations, inspirations and cause/effect relations. They are not indicated anywhere in the Apron Diagram. They may move up and down or anywhere, and they involve time.

The platform of the deep ecology movement is grounded in religion or philosophy. In a loose sense it may be said to be derived from the fundamentals. The situation only reminds us that a set of very similar or even identical conclusions may be drawn from divergent premises. The platform is the same, the fundamental premises differ. One must avoid looking for one definite philosophy or religion among the supporters of the deep ecology movement. Fortunately there is a rich manifold of fundamental views compatible with the platform of the deep ecology movement. Furthermore, there is a manifold of kinds of consequences derived from the platform.

We must take four levels into account: (1) verbalized fundamental philosophical and religious ideas and intuitions; (2) the platform of the deep ecology movement; (3) more or less general consequences derived from the platform—lifestyles and general policies of every kind; and (4) concrete situations and practical decisions made in them.

The possibility of the platform principles being derived from a plurality of mutually inconsistent premises—A-set and the B-set—is illustrated in the upper part of the Apron Diagram. A can be Buddhism, and B can be Christianity: or A may be Spinoza's philosophy, and B may be Ecosophy T. Similarly, the lower part of the diagram illustrates how, with one or more of the eight principles as part of a set of premises, mutually inconsistent conclusions may logically be derived, leading to the C-set and D-set of concrete decisions. C may be inspired by a sort of Christianity, and D by a sort of Buddhism: or, again, C may be Spinoza-inspired whilst D follows a certain ecological philosophy. (Unfortunately, the relation of deepness in the Apron Diagram leads upwards. To avoid mixing metaphors, the apron should be turned upside down.)

A distinction between the four levels is important. Supporters of the deep ecology movement have ultimate views from which they derive their acceptance of the platform, but those views may be very different from person to person, and from group to group. Likewise, supporters may disagree about what follows from the eight points, partly because they interpret them differently, partly because what follows does not follow from those eight points alone, but from a wider set of premises, and these may be in conflict.

The deep ecology movement thus can manifest both plurality and unity: unity at Level 2, and plurality at the other levels.

3

Self-Realization: An Ecological Approach to Being in the World

Arne Naess

For about 2500 years humankind has struggled with basic questions about who we are, what we are heading for, and what kind of reality we are part of. Two thousand, five hundred years is a short period in the lifetime of a species, and still less in the lifetime of the Earth, to whose surface we belong as mobile parts. I am not capable of saying very new things, but I can look at things from a *somewhat* different angle, using somewhat different conceptual tools and images.

What I am going to say more or less in my own way and that of my friends may roughly be condensed into the following six points:

1. We under-estimate ourselves. I emphasize 'self.' We tend to confuse it with the narrow ego.

2. Human nature is such that with sufficient allsided maturity we cannot avoid 'identifying' our self with all living beings, beautiful or ugly, big or small, sentient or not. I need of course to elucidate my concept of identifying. I'll come back to that.

The adjective 'allsided' in 'allsided maturity' deserves note: Descartes seemed to be rather immature in his relation to animals, Schopenhauer was not much advanced in his relation to family (kicking his mother down a staircase?), Heidegger was amateurish—to say the least—in his political behavior. Weak identification with non-humans is compatible with maturity in some major sets of relations, such as those towards family or friends. I use the qualification 'allsided,' that is, 'in all major relations.'

3. Traditionally the *maturity of self* has been considered to develop through three stages, from ego to social self, comprising the ego, and from there to the metaphysical self, comprising the social self. But Nature is then largely left out in the conception of this process. Our home, our immediate environment, where we belong as children, and the identification with human living beings, are largely ignored. I therefore tentatively introduce, perhaps for the first time ever, a concept of *ecological self*. We may be said to be in, of and for Nature from our very beginning. Society and human relations are important, but our self is richer in its constitutive relations. These relations are not only relations we have to other humans and the human community. (I have introduced the term 'mixed community' for communities where we consciously and deliberately live closely together with certain animals.)

4. Joy of life and meaning of life is increased through increased self-realization. That is, through the fulfillment of potentials each has, but which never are exactly the same for any pair of living beings. Whatever the differences, increased self-realization implies broadening and deepening of self.

5. Because of an inescapable process of identification with others, with growing maturity, the self is widened and deepened. We 'see ourselves in others.' Self-realization is hindered if the self-realization of others, with whom we identify, is hindered. Love of our self will fight this obstacle by assisting in the self-realization of others according to the formula 'live and let live!' Thus, all that can be achieved by altruism—the *dutiful, moral* consideration of others—can be achieved—and much more—through widening and deepening our self. Following Kant we then act beautifully, but neither morally nor immorally.

6. A great challenge of today is to save the planet from further devastation that violates both the enlightened self-interest of humans and non-humans, and decreases the potential of joyful existence for all.

Now, proceeding to elaborate these points, I shall start with the peculiar and fascinating terms 'ego,' 'self.'

The simplest answer to who and what I am is to point to my

body, using my finger. But clearly I cannot identify my self or even my ego with my body. Example: Compare "I know Mr. Smith" with "My body knows Mr. Smith"; "I like poetry" with "My body likes poetry"; "The only difference between us is that you are a Presbyterian and I am a Baptist," with "the only difference between our bodies is that your body is Presbyterian whereas mine is Baptist."

In the above sentences we cannot substitute 'my body' for 'I.' Nor can we substitute 'my mind' or 'my mind and body' for 'I.' More adequately we may substitute 'I as a person' for 'I,' but of course this does not tell us what the ego or self is.

A couple of thousand years of philosophical, psychological and social-psychological thinking has not brought us to any stable conception of the I, ego, or the self. In modern psychotherapy these notions play an indispensable role, but, of course, the practical goal of therapy does not necessitate philosophical clarification of the terms. It is for the purpose of this paper important to remind ourselves about the strange and marvelous phenomena we are dealing with. They are extremely close. Perhaps the very nearness of these objects of thought and reflection adds to our difficulties. I shall only offer one single sentence resembling a definition of the ecological self. The ecological self of a person is that with which this person identifies.

This key sentence (rather than definition) about the self shifts the burden of clarification from the term 'self' to that of identification, or rather 'process of identification.'

What would be a paradigm situation of identification? It is a situation in which identification elicits intense empathy. My standard example has to do with a non-human being I met 40 years ago. I looked through an old-fashioned microscope at the dramatic meeting of two drops of different chemicals. A flea jumped from a lemming strolling along the table and landed in the middle of the acid chemicals. To save it was impossible. It took many minutes for the flea to die. Its movements were dreadfully expressive. What I felt was, naturally, a painful compassion and empathy. But the empathy was not basic, it was the process of identification, that "I see myself in the flea." If I was alienated from the flea, not seeing intuitively anything even resembling myself, the death struggle would have left me indifferent. So there must be identification in order for there to be compassion and, among humans, solidarity.

One of the authors contributing admirably to clarification of the study of self is Erich Fromm.

The doctrine that love for oneself is identical with 'selfishness' and an alternative to love for others has pervaded theology, philosophy, and popular thought; the same doctrine has been rationalized in scientific language in Freud's theory of narcissism. Freud's concept presupposes a fixed amount of libido. In the infant, all of the libido has the child's own person as its objective, the stage of 'primary narcissism,' as Freud calls it. During the individual's development, the libido is shifted from one's own person toward other objects. If a person is blocked in his 'object-relationships,' the libido is withdrawn from the objects and returned to his or her own person; this is called 'secondary narcissism.' According to Freud, the more love I turn toward the outside world the less love is left for myself, and vice versa. He thus describes the phenomenon of love as an impoverishment of one's self-love because all libido is turned to an object outside oneself.[1]

What Erich Fromm attributes here to Freud, we today attribute to the shrinkage of self-perception implied in the ego-trip fascination. Fromm opposes such shrinkage. The following quotation concerns love of persons, but as 'ecosophers' we find the notions of 'care, respect, responsibility, knowledge' applicable to living beings in the wide sense.

Love of others and love for ourselves are not alternatives. On the contrary, an attitude of love toward themselves will be found in all those who are capable of loving others. Love, in principle, is indivisible as far as the connection between 'objects' and one's own self is concerned. Genuine love is an expression of productiveness and implies care, respect, responsibility, and knowledge. It is not an 'effect' in the sense of being effected by somebody, but an active striving for the growth and happiness of the loved person, rooted in one's own capacity to love.[2]

Fromm is very instructive about unselfishness—diametrically opposite to selfishness, but still based upon alienation and narrow perceptions of self. We might add that what he says applies also to persons experiencing sacrifice of themselves.

> The nature of unselfishness becomes particularly apparent in its effect on others and most frequently, in our culture, in the effect the 'unselfish' mother has on her children. She believes that by her unselfishness her children will experience what it means to be loved and to learn, in turn, what it means to love. The effect of her unselfishness, however, does not at all correspond to her expectations. The children do not show the happiness of persons who are convinced that they are loved; they are anxious, tense, afraid of the mother's disapproval, and anxious to live up to her expectations. Usually, they are affected by their mother's hidden hostility against life, which they sense rather than recognize, and eventually become imbued with it themselves. . . .
>
> If one has a chance to study the effect of a mother with genuine self-love, one can see that there is nothing more conducive to giving a child the experience of what love, joy, and happiness are than being loved by a mother who loves herself.[3]

We need an environmental ethics, but when people feel they unselfishly give up, even sacrifice, their interest in order to show love for Nature, this is probably in the long run a treacherous basis for conservation. Through identification they may come to see their own interest served by conservation, through genuine self-love, love of a widened and deepened self.

At this point the notion of a being's interest furnishes a bridge from self-love to self-realization. It should not surprise us that Erich Fromm, influenced as he is by Spinoza and William James, makes use of that bridge. What is considered to constitute self-interest, he asks, and he answers:

> There are two fundamentally different approaches to this problem. One is the objectivistic approach most clearly formulated by Spinoza. To him self-interest or the interest 'to seek one's profit' is identical with virtue.
>
> The more each person strives and is able to seek his profit, that is to say, to preserve his being, the more virtue does he possess; on the other hand, in so far as each person neglects his own profit

he is impotent. According to this view, the interest of humans is to preserve their existence, which is the same as realizing their inherent potentialities. This concept of self-interest is objectivistic inasmuch as 'interest' is not conceived in terms of the subjective feeling of what one's interest is but in terms of what the nature of a human is, 'objectively.'[4]

'Realizing inherent potentialities' is one of the good less-than-ten-word clarifications of 'self-realization.' The questions "Which are the inherent potentialities of the beings of species x?" and "Which are the inherent potentialities of this specimen y of species x?" obviously lead to reflections about and studies of x and y.

As humans we cannot just follow the impulses of the moment when asking for our inherent potentialities. It is something like this which Fromm means when calling an approach 'objectivistic,' opposing it to an approach 'in terms of subjective feeling.' Because of the high estimation of feeling and low estimate of so-called 'objectivization' (*Berdinglichung*, reification) within deep ecology, the terminology of Fromm is not adequate today, but what he means to say is appropriate. And it is obviously relevant when we deal with other species than humans: animals and plants have interests in the sense of ways of realizing inherent potentialities which we can only study interacting with them. We cannot rely on our momentary impulses, however important they are in general.

The expression 'preserve his being' in the quotation from Spinoza is better than 'preserve his existence' because the latter is often associated with physical survival and 'struggle for survival.' A still better translation is perhaps 'persevere in his being' *(perseverare in suo esse)*. It has to do with acting out one's own nature. To survive is only a necessary condition, not a sufficient condition.

The conception of self-realization as dependent upon our insight into our own potentialities makes it easy to see the possibility of ignorance and misunderstanding as to which are these potentialities. The ego-trip interpretation of the potentialities of humans presupposes a marked underestimation of the richness and broadness of our potentialities. In Fromm's terms, "man can deceive himself about his real self-interest if he is ignorant of his self and its real needs...."[5]

The 'everything hangs together' maxim of ecology applies to the self and its relation to other living beings, ecosystems, the ecosphere, and the Earth with its long history.

The scattered human habitation along the arctic coast of Norway is uneconomic, unprofitable, from the point of view of the current economic policy of our welfare state. The welfare norms require that every family should have a connection by telephone (in case of illness). This costs a considerable amount of money. The same holds for mail and other services. Local fisheries are largely uneconomic perhaps because a foreign armada of big trawlers of immense capacity is fishing just outside the fjords. The availability of jobs is crumbling.

The government, therefore, heavily subsidized the resettlement of people from the arctic wildernesses, concentrating them in so-called centers of development, that is, small areas with a town at the center. But the people, as persons, are clearly not the same when their bodies have been thus transported. The social, economic *and natural setting* is now vastly different. The objects with which they work and live are completely different. There is a consequent loss of personal identity. "Who am I?" they ask. Their self-respect, self-esteem is impaired. What is adequate in the so-called periphery of the country is different from what counts at the so-called centers.

If people are relocated or, rather, transplanted, from a steep mountainous place to a plain, they also realize, but too late, that their home-place has been part of themselves, they have identified with features of the place. And the way of life in the tiny locality, the density of social relations has formed their persons. Again, 'they are not the same as they were.'

Tragic cases can be seen in other parts of the Arctic. We all regret the fate of the Eskimos, their difficulties in finding *a new identity*, a new social and a new, more comprehensive ecological self. The Lapps of Arctic Norway have been hurt by interference with a river for the purpose of hydroelectricity. In court, accused of illegal demonstration at the river, one Lapp said that the part of the river in question was 'part of himself.' This kind of spontaneous answer is not uncommon among people. They have not heard about the philosophy of the wider and deeper self, but they talk spontaneously as if they had.

The sentence "This place is part of myself" we may try to make intellectually more understandable by reformulations, for example 'My relation to this place is part of myself.' 'If this place is destroyed something in me is destroyed.' 'My relation to this place is such that if the place is changed I am changed.'...

One drawback with these reformulations is that they make it easy to continue thinking of two completely separable, real entities, a self and the place, joined by an external relation. The original sentence, rather, conveys the impression that there is an internal relation of sorts. 'Of sorts' because we must take into account that it may not be reciprocal. If I am changed, even destroyed, the place would be destroyed according to one usual interpretation of 'internal relation.' From the point of phenomenology and the 'concrete content' view the reciprocity holds, but that is a special interpretation. We may use an interpretation such that if we are changed, the river need not be changed.

The reformulation 'If this place is destroyed something in me is killed' perhaps articulates some of the feelings usually felt when people see the destruction of places they deeply love or to which they have the intense feeling of belonging. Today more space is violently transformed per human being than ever, at the same time as their number increases. The kind of 'killing' referred to occurs all over the globe, but very rarely does it lead to strong counter-reaction. Resignation prevails. 'You cannot stop progress.'

The newborn lacks, of course, any conceptions, however rudimentary, corresponding to the tripartition—subject, object, medium. Probably the conception (not the concept) of one's own ego comes rather late, say after the first year. A vague net of relations comes first. This network of perceived and conceived relations is neutral, fitting what in British philosophy was called 'neutral monism.' The whole, their universe altogether, lacks the tripartition at this early stage. In a sense, it is this basic sort of crude monism we are working out anew, not by trying to be babies again, but by better understanding our ecological self. It has not had favorable conditions of development since before the time the Renaissance glorified our ego by putting it in some kind of opposition to what else there is.

What is now the practical importance of this conception of a wide and deep ecological self?

Defending Nature in our rich, industrial society, the argument of the opponent often is that we are doing it in order to secure beauty, recreation, sport, and other non-vital interests for us. It makes for strength if we, after honest reflection, find that we feel threatened in our innermost self. If so, we more convincingly defend a vital interest, not only something out there. We are engaged in self-defense. And to defend fundamental *human* rights is vital self-defense.

The best introduction to the psychology of the self is still to be found in the excellent and superbly readable book *Principles of Psychology*, published in 1890 by the American psychologist and philosopher William James. His 100-page chapter on the consciousness of self stresses the plurality of components of the wide and deep self as a complex entity. (Unfortunately he prefers to talk about the plurality of selves. I think it may be better to talk about the plurality of the components of the wide self.)

The plurality of components can be easily illustrated by reference to the dramatic phenomenon of alternating personality.

> Any man becomes, as we say *inconsistent* with himself if he forgets his engagements, pledges, knowledge and habits. . . . In the hypnotic trance we can easily produce an alternation of personality, . . . by telling him he is an altogether imaginary personage.[6]

If we say about someone that he or she is not him or her self today, we may refer to a great many different *relations* to other people, to material things and certainly, I maintain, to what we call his or her environment, the home, the garden, the neighborhood. . . .

When James says that these relations *belong* to the self, it is of course not in the sense that the self has eaten the home, the environment, etc. Such an interpretation testifies that the self is still identified with the body. Nor does it mean that an *image* of the house inside the consciousness of the person belongs to the self. When somebody says about a part of a river-landscape that it is part of himself, we intuitively grasp roughly what he means. But it is of course difficult to elucidate the meaning in philosophical or psychological terminology.

A last example taken from William James: We understand what is meant when somebody says "As a man I pity you, but as an official I must show you no mercy." Obviously the self of an official cannot empirically be defined except as relations in a complex social setting. Thus, the self cannot possibly be inside the body, or inside a consciousness. Enough! The main point is that we do not hesitate *today*, being inspired by ecology and a revived intimate relation to Nature, to recognize and accept wholeheartedly our ecological self.

The next section is rather metaphysical. I do not *defend* all the views presented in this part of my paper. I wish primarily to inform you about them. As a student and admirer since 1930 of Gandhi's non-violent direct actions in bloody conflict, I am inevitably influenced by his metaphysics, which to him personally furnished tremendously powerful motivation and which contributed to keeping him going until his death. His supreme aim was not India's *political* liberation. He led a crusade against extreme poverty, caste suppression, and against terror in the name of religion. This crusade was necessary, but the liberation of the individual human being was his supreme aim. It is strange for many to listen to what he himself said about this ultimate goal:[7]

> What I want to achieve—what I have been striving and pining to achieve these thirty years—is self-realization, to see God face to face, to attain *Muksha* (Liberation). I live and move and have my being in pursuit of that goal. All that I do by way of speaking and writing, and all my ventures in the political field, are directed to this same end.

This sounds individualistic to the Western mind. A common misunderstanding. If the self Gandhi is speaking about were the ego or the 'narrow' self ('jiva') of egocentric interest, the 'ego-trips,' why then work for the poor? It is for him the supreme or universal Self— the *atman*—that is to be realized. Paradoxically, it seems, he tries to reach self-realization through 'selfless action,' that is, through reduction of the dominance of the narrow self or the ego. Through the wider Self every living being is connected intimately, and from this intimacy follows the capacity of *identification* and as its natural consequences, the practice of non-violence. No moralizing is needed, just as we need no morals to make us breathe. We need to cultivate our insight:

> The rockbottom foundation of the technique for achieving the power of non-violence is belief in the essential oneness of all life.

Historically we have seen how Nature conservation is non-violent at its very core. Gandhi says:

> I believe in *advaita* (non-duality), I believe in the essential unity of man and, for that matter, of all that lives. Therefore I believe that if one man gains spirituality, the whole world gains with him and, if one man fails, the whole world fails to that extent.

Surprisingly enough, Gandhi was extreme in his personal consideration for the self-realization of other living beings than humans. When travelling he brought a goat with him to satisfy his need for milk. This was part of a non-violent demonstration against certain cruel features in Hindu ways of milking cows. Furthermore, some European companions who lived with Gandhi in his ashrams were taken aback that he let snakes, scorpions and spiders move unhindered into their bedrooms—animals fulfilling their lives. He even prohibited people from having a stock of medicines against poisonous bites. He believed in the possibility of satisfactory coexistence and he proved right. There were no accidents. Ashram people would naturally look into their shoes for scorpions before using them. Even when moving over the floor in darkness one could easily avoid trampling on one's fellow beings. Thus, Gandhi recognized a basic, common right to live and blossom, to self-realization in a wide sense applicable to any being that can be said to have interests or needs.

Gandhi made manifest the internal relations between self-realization, non-violence and what sometimes has been called biospherical egalitarianism.

In the environment in which I grew up, I heard that what is serious in life is to get to be somebody—to outdo others in something, being victorious in comparison of abilities. What today makes this conception of the meaning and goal of life especially dangerous is the vast international economic competition. Free market, perhaps, yes, but the law of demand and supply of separate, isolatable 'goods and services,' independent of needs, must not be made to reign over increasing areas of our life.

Ability to cooperate, to work with people, making them feel good *pays*, of course, in a fiercely individualist society, and high positions may require that; but only as long as ultimately it is subordinated to career, to the basic norms of the ego-trip, not to a self-realization worth the name.

To identify self-realization with the ego-trip manifests a vast underestimation of the human self.

According to the usual translation of Pali or Sanskrit texts, Buddha taught his disciples that the human mind should embrace all living things as a mother cares for her son, her only son. Some of you who never would feel it meaningful or possible that a human *self* could embrace all living things might stick to the usual translation. We shall then only ask that your *mind* embrace all living beings, and that you realize your good intention to care and feel with compassion.

If the Sanskrit word is *atman*, it is instructive to note that this term has the basic meaning of 'self,' rather than 'mind' or 'spirit,' as you see in translations. The superiority of the translation using the word 'self' stems from the consideration that if your 'self' in the wide sense embraces another being, you need no moral exhortation to show care. Surely you care for yourself without feeling any moral pressure to do it—provided you have not succumbed to a neurosis of some kind, developing self-destructive tendencies, or hating yourself.

Incidentally, the Australian ecological feminist Patsy Hallen [see her article in this anthology, eds.] uses a formula close to that of Buddha: We are here to embrace rather than conquer the world. It is of interest to notice that the term 'world' is here used rather than 'living beings.' I suspect that our thinking need not proceed from the notion of living being to that of the world, but we will conceive reality or the world we live in as alive in a wide, not easily defined sense. There will then be no non-living beings to care for.

If self-realization or self-fulfillment is today habitually associated with life-long ego-trips, isn't it stupid to use this term for self-realization in the widely different sense of Gandhi, or, less religiously loaded, as a term for widening and deepening your 'self' so it embraces all life forms? Perhaps it is. But I think the very popularity of the term makes people listen for a moment, feeling safe. In that moment the notion of a greater 'self' should be introduced, contending that if they equate self-realization with ego-trips, they seriously *underestimate* themselves. "You are much greater, deeper, more generous and capable of more dignity and joy than you think! A wealth of non-competitive joys is open to you!"

But I have another important reason for inviting people to think in terms of deepening and widening their selves, *starting* with the ego-trip as a crudest, but inescapable point zero. It has to do with a notion usually placed as the opposite of the egoism of the ego-trip, namely the notion of *altruism*. The Latin term *ego* has as its opposite the *alter*. Altruism implies that *ego* sacrifices its interest in favor of the other, the *alter*. The motivation is primarily that of duty: it is said that we *ought* to love others as strongly as we love our self.

It is, unfortunately, very limited what humankind is capable of loving from mere duty, or, more generally, from moral exhortation. From the Renaissance to the Second World War about 400 cruel wars were fought by Christian nations for the flimsiest of reasons. It seems to me that in the future more emphasis has to be given to the conditions under which we most naturally widen and deepen our 'self.' With a sufficiently wide and deep 'self,' ego and alter as opposites are stage by stage eliminated. The distinction is in a way transcended.

Early in life, the social 'self' is sufficiently developed so that we do not prefer to eat a big cake alone. We share the cake with our friends and nearest relations. We identify with these people sufficiently to see our joy in their joy, and see our disappointment in theirs.

Now it is the time *to share* with all life on our maltreated Earth through the deepening identification with life forms and the greater units, the ecosystems, and Gaia, the fabulous, old planet of ours.

Immanuel Kant introduced a pair of contrasting concepts which deserve to be extensively used in our effort to live harmoniously in, for and of Nature: The concept of 'moral act' and that of 'beautiful act.'

Moral acts are acts motivated by the intention to follow the moral laws, at whatever the cost, that is, to do our moral duty solely out of respect for that duty. Therefore, the supreme *test* of our success in performing a pure, moral act is that we do it completely against our inclination; that we, so to say, hate to do it, but are compelled by our respect for the moral law. Kant was deeply awed by two phenomena, "the heaven with its stars above me and the moral law within me."

But if we do something we should do according to a moral law, but do it out of inclination and with pleasure—what then? Should we then abstain or try to work up some displeasure? Not at all, according to Kant. If we do what morals say is right because of positive inclination, then we perform a *beautiful* act. Now, my point is that perhaps we should in environmental affairs primarily try to influence people towards beautiful acts. Work on their inclinations rather than morals. Unhappily, the extensive moralizing within environmentalism has given the public the false impression that we primarily ask them to sacrifice, to show more responsibility, more concern, better morals. As I see it we need the immense variety of sources of joy opened through increased sensitivity towards the richness and diversity of life, landscapes of free Nature. We all can contribute to this individually, but it is also a question of politics, local and global. Part of the joy stems from the consciousness of our intimate relation to something bigger than our ego, something which has endured through millions of years and is worth continued life for millions of years. The requisite care flows naturally if the 'self' is widened and deepened so that protection of free Nature is felt and conceived as protection of ourselves.

Academically speaking, what I suggest is the supremacy of environmental ontology and realism over environmental ethics as a means of invigorating the environmental movement in the years to come. If reality is like it is experienced by the ecological self, our behavior *naturally* and beautifully follows norms of strict environmental ethics.

We certainly need to hear about our ethical shortcomings from time to time, but we more easily change through encouragement and through deepened perception of reality and our self. That is, deepened realism. How is that to be brought about? The question needs to be treated in another paper! It is more a question of community therapy than community science: Healing our relations to the widest community, that of all living beings.

The subtitle of this paper is "An Ecological Approach to Being in the World." I am now going to speak a little about 'Nature' with all the qualities we spontaneously experience, as identical with the reality we live in. That means a movement from being in the world to being in Nature. Then, at last, I shall ask for the goal or purpose of being in the world.

Is joy in the subject? I would say "no." It is just as much or little in the object. The joy of a joyful tree is primarily "in" the tree, we should say—if we are pressed to make a choice between the two possibilities. But we should not be pressed. There is a third position. The joy is a feature of the *indivisible,* concrete unit of subject, object and medium. In a sense self-realization involves experiences of the infinitely rich joyful aspect of reality. It is misleading, according to my intuitions, to locate joys inside my consciousness. What is joyful is something that is not 'subjective,' it is an attribute of a reality wider than a conscious ego. This is philosophically how I contribute to the explanation of the internal relation between joy, happiness, and human self-realization. But this conceptual exercise is mainly of interest to an academic philosopher. What I am driving at is probably something that may be suggested with less conceptual gymnastics: It is unwarranted to believe that how we feel Nature to be is not how Nature really is. It is rather that reality is so rich that we cannot see everything at once, but separate parts or aspects of separate moods. The joyful tree I see in the morning light is not the sorrowful one I see in the night, even if they in abstract structure (physically) are the same.

It is very human to ask for the ultimate goal or purpose for being in the world. This may be a misleading way of putting a question. It may seem to suggest that the goal or purpose must be somehow outside or beyond the world. Perhaps this can be avoided by living out "in the world." It is characteristic for our time that we subjectivize and individualize the question asked of each one of us: What do *you* consider to be the ultimate goal or purpose of *your* life? Or, we leave out the question of priorities and ask simply for goals and purposes.

The main title of this paper is partly motivated by the conviction that 'self-realization' is an adequate key-term expression one uses to answer the question of ultimate goal. It is of course only a key-term. An answer by a philosopher can scarcely be shorter than the little book *Ethics* by Spinoza.

In order to understand the function of the term 'self-realization' in this capacity, it is useful to compare it to two others, 'pleasure' and 'happiness.' The first suggests hedonism, the second eudaemonism in a professional philosophical, but just as vague and ambiguous jargon. Both terms connote states of feeling in a broad sense of the term. Having pleasure or being happy is to *feel* well. One may of course find the term happiness to connote something different from this, but in the way I use 'happiness,' one standard set of replies to the question "How do you feel?" is "I feel happy" or "I feel unhappy." This set of answers would be rather awkward: "I feel self-realized" or "I do not feel self-realized."

The most important feature of self-realization as compared to pleasure and happiness is its dependence upon a view of human capacities, better potentialities. This again implies a view of what is human nature. In practice it does not imply a general doctrine of human nature. That is the work of philosophical fields of research. An individual whose attitudes are such that I would say that he or she takes self-realization as the ultimate or fundamental goal has to have a view of his or her nature and potentialities. The more they are realized the more there is self-realization. The question "How do you feel?" may be honestly answered in the positive or negative whatever the level of self-realization. The question may, in principle, be answered in the negative, but at the point following Spinoza I take the valid way of answering to be positive. The realization of fulfillment—using a somewhat less philosophical jargon—of the potentialities of oneself is *internally* related to happiness, but not in such a way that *looking* for happiness you realize yourself. This is a clear point, incidentally, in John Stuart Mill's philosophy. You should not look hard for happiness. That is a bad way even if you take, as Mill does, happiness as the ultimate or fundamental goal in life. I think that to look for self-realization is a better way. That is, to develop your capacities—using a rather dangerous word because it is easily interpreted in the direction of interpersonal, not intrapersonal, competition. But even the striving implied in the latter term may mislead. Dwelling in situations of intrinsic value, spontaneous non-directed awareness, relaxing from striving, is conducive to self-realization as I understand it. But there are, of course, infinite variations among humans according to cultural, social, individual differences. This makes the key-term self-realization abstract in its generality. But nothing more can be expected when the question is posed like it is: What might deserve the name of ultimate or fundamental goal? We may reject the meaningfulness of such a question—I don't—but for us for whom it has meaning, the answer using few words is bound to be abstract and general.

Going back to the triple key-terms pleasure, happiness, self-realization, the third has the merit of being clearly and forcefully applicable to any beings with a specific range of potentialities. I limit the range to living beings, using 'living' in a rather broad sense. The terms 'pleasure' and 'happiness' I do not feel are so easily generalized. With the rather general concept of 'ecological self' already introduced, the concept of self-realization naturally follows. Let us consider the praying mantis, the formidable group of voracious insects. They have a nature fascinating to many people. Mating is part of their self-realization, but some males are eaten when performing the act of copulation. Is he happy, is he having pleasure? We don't know. Well done if he does! Actually he feeds his partner so that she gets strong offspring. But it does not make sense to me to attribute happiness to these males. Self-realization yes, happiness no. I maintain the internal relation between self-realization and happiness among people and among some animal groups. As a professional philosopher I am tempted to add a point where I am inspired by Zen Buddhism and

Spinoza: Happiness is a feeling yes, but the act of realizing a potential is always an interaction involving as one single concrete unit, one gestalt as I would say, three abstract aspects, subject, object, medium. What I said about joyfulness in Nature holds of happiness in Nature. We should not conceive them as mere subjective feelings.

The rich reality is getting even richer through our specific human endowments; we are the first kind of living beings we know of which have the potentialities of living in community with all other living beings. It is our hope that all those potentialities will be realized—if not in the near future, at least in the somewhat more remote future.

Notes

1. Clark E. Moustakas, ed., *The Self: Explorations in Personal Growth,* New York, Harper Books, 1956, p. 58. This and the following quotations are from Fromm's contribution "Selfishness, Self-love, and Self-interest."

2. Ibid., p. 59.

3. Ibid., p. 62.

4. Ibid., p. 63.

5. Ibid.

6. James, Vol. 1, p. 379.

7. This and other quotations from Gandhi are taken from my *Gandhi and Group Conflict,* Oslo, 1974, p. 35, where the metaphysics of self-realization is treated more thoroughly.

4

The Systematization of the Logically Ultimate Norms and Hypotheses of Ecosophy T

Arne Naess

(a) The idea of models of logical relations

The complete formulation of an ecosophy is out of the question: the complexity and flexibility of such a living structure make that impossible, perhaps even meaningless. There may also be logical reasons for the impossibility of the formulating of a total view: it would be like a gestalt without a background, an absurdity. One may, however, simulate such a system. One may make a model of parts of it, isolating certain patterns and aspects of it for close scrutiny, implicitly pretending that the rest somehow exists in the realm of pure thought.

In what follows I shall work out such a model. It expresses the vision of an ecosophy in the form of a pyramid or tree.

The direction from top to bottom, from theory to praxis, is one of logical, not genetic or historical, derivation. It is not a ranking order. It does not indicate value priorities. At the top levels there is a small number of general and abstract formulations, at the bottom singular and concrete ones, adapted to special situations, communities, time intervals, and actions.

The direction from the bottom up offers the genetic and historical derivation—including all the motivations and impulses resulting in formulations of norms and hypotheses.

What is modeled is a moving, ever-changing phenomenon: norms and hypotheses being derived more or less logically, applied in praxis, and the outcome motivating changes. The tree can be arranged to form a triangle or parallelogram with a wide horizontal base line and a narrow top line. The difference in breadth expresses the fact that from the abstract and general norms and hypotheses indefinitely many more specialized norms and hypotheses follow, giving rise to indefinitely many decisions in concrete situations.

If we decide to reject a low-level norm, this implies that we will have to modify some hypotheses or norms at higher levels. The whole upper pyramid gets to be shaky. However, a rejection tends, in practice, to cause only slight modification, or simply the adoption of a somewhat different precisation [a technical term introduced by Naess] of a higher norm or hypothesis *formulation*.

A sentence like 'Seek Self-realization!' will be interpreted within social science as a sentence in the imperative mood, and from a social point of view it is pertinent to ask: who are the 'senders' and who are the 'receivers'? Not necessarily so in T_0-expositions of models of normative systems. [T_0 stands for the least precise and most general level of articulation. A T_0-level sentence allows all the possible interpretations and directions of precisation. A sentence T_1 is more precise than T_0 in that T_0 permits all the interpretations of T_1, while T_1 does not admit all the interpretations of T_0.] The question of how to understand the function of a one-word sentence like 'Self-realization!' is a large and deep one which I am not going to attack. Let it suffice to say that there are examples of the use of the exclamation mark with a sender, but no definite receiver. When terrible things happen, like the collapsing of a bridge, or the loss of one's keys down a drain, we may meaningfully say 'No!,' but at least some of us have in such cases no definite receiver group in mind. The function of the sentence is clear enough without. An archetype of this function appears in the Bible when God says 'Let there be light!' To ask who are the intended receivers of the exclamation is in this case rather intriguing if we suppose there is not yet anything created. But to ask the question is itself questionable.

In short I find it to be my duty to point out that there are questions concerning the function of norm-sentences in the exposition at hand which one should acknowledge but not necessarily engage in 'solving.'

What follows is only one particular exposition of Ecosophy T. Other versions may be cognitively equivalent, expressing the same concrete content in a different abstract structure.

(b) Formulation of the most basic norms and hypotheses

N1: Self-realization!

H1: The higher the Self-realization attained by anyone, the broader and deeper the identification with others.

H2: The higher the level of Self-realization attained by anyone, the more its further increase depends upon the Self-realization of others.

H3: Complete Self-realization of anyone depends on that of all.

N2: Self-realization for all living beings!

Comments:

The four formulations N1, H1, H2, and H3 make up the first level of the survey. N1 and H1 are ultimates in the sense of not being derivable from the others within the chosen version of the logical systematization of Ecosophy T. H2, H3, and N2 are supposed to be logically derivable from the ultimates. Formal rigor would of course require us to add some premises which would be of greater interest to the logician than to the ecosopher. (For instance: if A identifies with B, and both are beings such that it is meaningful to talk about their higher Self-realization levels, complete Self-realization of A requires complete Self-realization of B. Our consolation: the formal logical derivation of the theorems of the first part of Spinoza's *Ethics* seems to require about 160 additional premises—but with them at hand consistency is achieved.)

All norm- and hypothesis-formulations are T_0-formulations, that is, at the most primitive level from the point of view of preciseness. The decrease of egocentricity is inevitably linked to an increase of identification and care for others. Which 'others'? One good answer is to draw circles of interest and care, corresponding to stages of development: family, clan, tribe, humanity. But obviously animals, especially the tamed or domestic, often enjoy interest, care, and respect (at times status of divinity) before humanity at large. The series of circles will differ in different cultures. In any case higher levels of realization of potentials of the self favor the Self-realization of others.

Considering the widening scope of identification as internally related to increased Self-realization, this increase depends on the Self-realization of others. This gives us H1. It implies that 'the others' do not lose their individuality. Here we stumble upon the old metaphysical set of problems of 'unity in diversity.' When the human being A identifies with B, and the wider self of A comes to comprise B, A is not supposed to reject the individuality of B. Thus, if B and A are persons, the self of A comprises that of B and vice versa.

The importance of H1 for the whole *conceptual* development of Ecosophy T stems from the way those who think it is a tenable hypothesis—those who feel at home with it—are apt to view nature and what is going on in nature. They see a lonely, desperately hungry wolf attacking an elk, wounding it mortally but being incapable of killing it. The elk dies after protracted, severe pains, while the wolf dies slowly of hunger. Impossible not to identify with and somehow feel the pains of both! But the nature of the conditions of life at least in our time is such that nothing can be done about the 'cruel' fate of both. The general situation elicits sorrow and the search for means to interfere with natural processes on behalf of any being in a state of panic and desperation, protracted pain, severe suppression or abject slavery. But this attitude implies that we deplore much that actually goes on in nature, that we deplore much that seems essential to life on Earth. In short, the assertion of H1 reflects an attitude opposed to any unconditional *Verherrlichung* [glorification] of life, and therefore of nature in general.

For H3, a somewhat more precise formulation would be 'Complete Self-realization of anyone depends on that of all beings which *in principle are capable of Self-realization*.' For the sake of brevity in the survey these beings are in what follows called 'living beings.' We define 'living beings' in this way.

The fact that N2 is derivable from N1 through the aforementioned hypotheses does not automatically make it into a purely instrumental norm in Ecosophy T. It is only instrumental *in relation* to

34

N1. A norm is purely instrumental only if its definition excludes it from being non-instrumental in any single relation. Example: it may pay in the long run to be honest, but this does not exclude the possibility that honesty can be a valid non-instrumental norm, independent of profits.

Saying unconditionally yes to N1 implies a yes to the question of whether Self-realization is something and something of value. Since there is nothing which could make Self-realization a purely instrumental norm, the yes is announcing its intrinsic value. Saying yes to N2 implies the intrinsic valuation of all living beings. From these two norms, and norms derived from them (plus hypotheses), the proposed formulations of the platform of the deep ecological movement are derivable. In a common philosophical terminology, the platform is expressing an axiology whereas Ecosophy T expresses a deontology. The latter classification is suspicious, however, because the exclamation mark of N1 does not imply that what is expressed is a communication to somebody. N1 rather expresses an ontology than deontology. It is, however, not the aim here to go far into professional philosophy.

(c) Norms and hypotheses originating in ecology

H4: Diversity of life increases Self-realization potentials.
N3: Diversity of life!
H5: Complexity of life increases Self-realization potentials.
N4: Complexity!
H6: Life resources of the Earth are limited.
H7: Symbiosis maximizes Self-realization potentials under conditions of limited resources.
N5: Symbiosis!

Comments:

These seven formulations make up the second part of the survey. Whereas the first level is squarely metaphysical, the second level is biologically colored, but still metaphysical because of the use of capital S in 'Self-realization.' More precise formulations would refer to general ecology and conservation biology rather than human ecology.

H4 introduces the central term of 'Self-realization *potential*.' In psychology and sociology there is much discussion of the potentials, potentialities, or possibilities which an individual, group, or institution has in life, including the life of nations. There is within ethics talk about talents and capacities and how to develop them. The term 'Self-realization' is a kind of generalization of this, except that, in using the capital letter S, certain norms are proclaimed which narrow down the range of what constitutes an increase of Self-realization.

H4 has a metaphysical background. Life is viewed as a kind of vast whole. The variety of forms of life, with their different capacities, realize, that is, bring into actuality, something which adds to that whole. They realize the Self-realization *potentials*. Each individual contains indefinitely many of these, not only one. An increase in qualitative diversity of life forms increases the possibility of potentials. From H4 and N1 therefore follows N3.

(d) The meaning of diversity, complexity, and symbiosis in the context of Self-realization

'Self-realization!' with a capital S is a norm formulation inspired by the part of philosophy traditionally called metaphysics. The terms diversity, complexity, and symbiosis are all borrowed from ecology. There is a resulting kind of terminological tension between the first two levels of the survey, as well as a general tension between 'the one and the many.'

The conceptual bridge from Self-realization to a positive evaluation of diversity, complexity, and symbiosis is furnished by a concept of Self-realization potentials, and the idea that the overall Self-realization in our world is increased by the realization of such potentials. (The realization is analogous to negative entropy.) No single being can completely realize the goal. The plural of potentials is crucial: it introduces plurality into unity. The intuition pushing us towards the 'Self' does not immediately acknowledge this.

A closely related idea is that of microcosm mirroring macrocosm, an idea especially potent during the Renaissance and now partly revived in hologram thinking. Each flower, each natural entity with the character of a whole (a gestalt) *somehow* mirrors or expresses the supreme whole. I say 'somehow' because I do not know of any good analysis of what is called mirroring here. The microcosm is not apart from the whole; the relation is not like that between a big elephant and a small mouse. Microcosm is essential for the existence of macrocosm. Spinoza was influenced by the idea when demanding an immanent God, not a God apart. The door is open for positive evaluation of an increase of the realization of potentialities, that is, of the possibility that more potentialities will be realized. This is meant to imply *continued evolution at all levels*, including protozoans, landscapes, and human cultures.

The realizations should be *qualitatively different*. Numerical abundance as such does not count. One way of emphasizing this distinction is to distinguish diversity from (mere) plurality. The term 'diversity' is well established in biology, mostly used in talking about diversity of species or of other qualitatively different living beings.

Further elaboration of the conception of diversity and the introduction of the concepts of complexity and symbiosis clearly require the support of hypotheses about the kind of universe we live in. Such support was, strictly speaking, necessary even when starting to talk about Self-realization, but only now is explicit mention of such support clearly needed. The universe which we shall limit ourselves to mentioning is our planet, the Earth, which we may also call 'Gaia' to emphasize its status as a living being in the widest sense.

I make a lot of implicit assumptions about the life conditions of Earth, especially its limitations. Any total view requires that.

Diversity may be defined so as to be only a *necessary* condition in the growth of realizing Self-realization potentialities. Then 'maximum diversity!' does not make sense, because many differences may not involve Self-realization and may be inconsistent with symbiosis. Better to imply qualitative difference as mentioned above to introduce concepts of difference which distinguish it from mere plurality. The ambivalence of plurality stems from finiteness—not only of our planet as a whole. However, the adjective 'maximum' is added to some expressions of Ecosophy T when diversity is introduced. The intention is to proclaim that there is no inherent limit to the positive character of growth of diversity. It is not intended that an increase is good even if it reduces the conditions for realizing other norms. If the adjective 'maximum' is to be retained, it must, at a more precise (T_I) level, be taken as an abbreviation for 'maximum, without hindering the realization of other norms in the system.' The presence of a norm of 'symbiosis!' in the system should re-emphasize this—it knits the bond between complexity and diversity.

Now let us turn to complexity.

If we are permitted to vary three factors a, b, c in spatial horizontal arrangements, we can only realize six different patterns: abc, acb, bac, bca, cab, cba. If we add one more basic factor, d, the number of arrangements increases to 'four factorial,' 24. This illustrates the *intimate relation between complexity and diversity*. When the number of elements increases linearly, the number of possible relationships increases factorially.

Let us then think that abc is a pattern of life, conceived as a kind of organismic or personal life. The pattern is characterized by three main functions or dimensions, a, b, c working together as a highly integrated system abc. Let the other five arrangements of a, b and c symbolize five other systems with the same number of dimensions.

The principle of self-preservation now may be said to consist minimally in an internal mechanism such that the system defends itself against reduction to 2-, 1-, or zero-dimensional ones, and also against transitions to systems symbolized through the other five patterns, *and* tends positively to develop into systems with more dimensions, thus more diversity and more complexity.

Complexity as opposed to complication is in Ecosophy T a quality of organisms and their relation to their environment. It is characterized by intimate interrelations, deep interdependence of a manifold of factors or elements. After death a rhinoceros as a breathing entity is no more, but it remains a tremendously complicated part of nature inhabited and invaded by millions of other, less complex organisms. A human victim of African sleeping sickness manifests the intimate interrelations between a human individual and colonies of the flagellate *Trypanosoma gambiensis*. Each of the flagellates has an unfathomable complexity of structure, but we recognize the human being as a still higher order of complexity.

If complexity is defined in the biological direction of the opposite of simplicity, 'maximum complexity!' cannot support Self-realization. Only if, as in the case of 'diversity!,' some restraining clause is inserted, could maximizing make sense.

Since the great time of the reptiles, limbs much more complex than the human hand have developed. The simplicity of the human hand is from this point of view a combined victory of simplicity and effectiveness over complexity. There should be no cult of complexity.

In biological texts colored by the conception of lower and higher animals, the term complexity nearly always is used in descriptions of *advantageous* cases of increases in complexity. 'Higher' functions are made possible through certain more complex differentiations of tissues. Eyes are developed from an earlier homogenous surface of skin. Less is said about unsuccessful increase of complexity, presumably because only species of great stability through millions of years have left fossils for us to study. I think it is most fruitful to use the term complexity as a rather general term covering also cases of no obvious advantage of any kind.

A simple biological example of increasing complexity of 'advanced' forms: the least complex type of sponge is similar to a sac. At one end there is an opening through which water and waste are thrown out. Through small openings in the walls water is drawn in. More complex forms have folded walls so that their surface is greater compared to the volume of the sac. This is thought to be an advance because it is a plus to have more surface cells compared to the number of other cells. A higher level of complexity is reached when special structures insure that waste is thrown out further away so that the sponge does not risk inhaling some of the waste again and again. On the whole zoologists are sure that increases of complexity have functions that could not be realized without those increases. There is no positive value to be attached to complexity as such, for instance walls of unequal thickness satisfying a certain rhythm, but of neither positive nor negative value for any discernible function of the organism.

In a diabolic world, evolution might have proceeded in many ways as in ours, except that parasitism might have made every being capable of conscious pain, suffering from birth to death. The increase of the amount and the intimacy of interrelations and interdependencies might, in the hypothetical world of diabolic parasitism, have resulted in a hellish level of intensity of suffering. Therefore complexity of organisms as such and complexity of interdependencies cannot in Ecosophy T be good in themselves.

From the point of view of biology, complexity comprises behavior and gestalt processes whereby increasing complexity of consciously experienced wholes can be realized. But also here mere complexity as such cannot yield an increase of Self-realization. The concept of symbiosis—life together—enters the framework. The existence of interdependencies in which all partners in a relationship are enriched furnishes a crucial idea in addition to diversity and complexity.

Proceeding from non-human to human ecology, the symbiosis idea may be illustrated in relation to various ways of realizing a caste system. When Gandhi sometimes spoke positively about a caste system, he had an ideal system in mind. Parents were to instruct children and work together with them as they grew up. No schools. The useful occupation of each family would be interrelated with and interdependent with families specializing in other kinds of services in the total community. Interaction between castes of this kind was to be encouraged, not prohibited. The status of each in the sense of dignity, respect, material standard of living, should be the same— an egalitarianism among castes, an illustration of symbiosis between groups in a community. Gandhi detested the actual state of affairs in the existing caste system in India. It certainly violated the norm of symbiosis.

In any kind of community we know of, there have been conflict and strife, in varying degrees. The norms of Ecosophy T are guidelines, and if elaborated into a comprehensive system would have to include norms for conflict solution. It is unrealistic to foresee full termination of deep group conflicts or even to wish such termination. The conditions of life on Earth are such that increase of Self-realization is dependent upon conflicts. What counts is the gradual increase of the status and application of nonviolence in group conflicts.

The *codification* of Ecosophy T is an action within the context of a conflict; it is my belief that many of the regrettable decisions in environmental conflicts in Norway and other places are made in a state of philosophical stupor. In that state people in power confuse narrow, superficial goals with fundamental broad goals derived from fundamental norms.

(e) Derivation of the norms of the local community

The next ecosophical principles to be incorporated are those of self-sufficiency, decentralization, and autonomy. These social principles are first to be linked to their biological counterparts.

The maximum success of Self-realization is realized through a certain balance of interactions between organisms and environment. The stimuli are not to be too erratic and not too monotonous. The organs of control must not completely dominate influences from the outside nor get overwhelmed. The limited possibilities of control make it, on the whole, important to have a fairly high degree of control of the spatially (personal) near environment, or the environment in which the basic needs are satisfied. If a basic need is only met through a many-stage interaction with remote areas, there are likely to be more forms of erratic obstacles, more dangers of being cut out through processes of chance character.

Let this be illustrated with the life-space models of the kind gestalt psychologist Kurt Lewin made use of (Figure 1).

Let A represent a living being in a two-dimensional space having four vital needs to satisfy. If the immediate environment furnishes, at least normally, satisfaction of the four needs, A can limit itself to try to control remote areas *only* if something unusual happens to the nearest. The quadruple $a1/1$ to $a1/4$ symbolizes the four sources of need satisfaction.

If the sources are $a2/1$, $a2/3$, $a2/5$, $a2/7$ and separated from A by interposed, qualitatively different parts $a1/1$ to $a1/4$ of the environment, the organism is vitally and normally dependent upon control

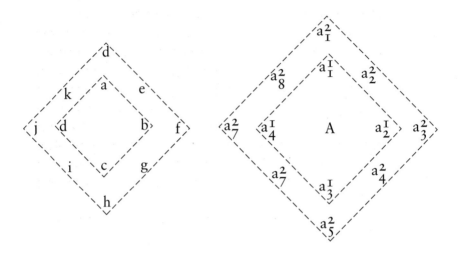

Figure 1

of these parts and also of a2/2, a2/4, a2/6, a2/8, the parts adjacent to the sources with another set of qualitatively different properties.

The illustration shows how the requirement of control increases with the remoteness of sources of satisfaction of needs—remoteness being measured in terms of distances in life space, not in kilometers. Making the supposition of limited means of control, the increase of remoteness correlates with increase of dangers, of inadequacy of powers of self-preservation and therefore with decrease of Self-realization potentials. By the degree of local self-sufficiency and autonomy we shall understand the degree to which the living being has its sources of basic need satisfactions, or more generally sources of Self-realization, nearby in the life space and, secondly, to what degree the organism has adequate control of this area to satisfy its needs.

The above model has been introduced with single living beings, especially persons, as units of life. This is didactically sound as long as it has no scientific pretensions. The same model is useful if taking collectives, communities, neighborhoods, societies, tribes—as units of life. But in that case we clearly need a model illustrating the relations within the collectives as well. Here we will not go into this.

By definition, single persons have less than maximum control over decisions of centralized authority, possibilities of control approach zero. The greater the manifold of persons and situations to be controlled, the greater the number of levels needed. Further, the greater the number of qualitatively different functions which are controlled, the more rapidly will the control by single persons tend towards zero. Centralization is here intended to be defined through the above factors.

Using the reasoning suggested above, a set of hypotheses and norms is proposed for Ecosophy T:

H8: Local self-sufficiency and cooperation favors increase of Self-realization.

H9: Local autonomy increases the chances of maintaining local self-sufficiency.

H10: Centralization decreases local self-sufficiency and autonomy.

N6: Local self-sufficiency and cooperation!

N7: Local autonomy!

N8: No centralization!

Comments:

Doubt No. 1: Does not the realization beyond a certain point of the three norms N6, N7, and N8 interpreted individualistically lead to strange conditions of life, in some ways similar to the famous terrifying 'state of nature' in the political philosophy of Thomas Hobbes?

Doubt No. 2: Do the lessons of ecology really support the norms? Rejecting the individualistic interpretation, we are therefore confronted with the difficult task of making them more precise with the help of other justifications, taking into account a serious concern of both individuals and collectives.

(f) Minimum conditions and justice: classes; exploitation

Human beings have needs. Any global policy of ecological harmony must distinguish the needs from mere wishes, that is to say from wishes that do not directly relate to a need.

Biological are those needs which must be unconditionally satisfied in order for an individual or species to survive. A minimum formula runs 'food, water, territory.' Then there are needs which are not necessary for all species. Clothing and some kind of shelter are necessary for most human groups, but not other species.

Further: we have needs necessary according to basic social organizations. We now approach needs which can only be separated from mere wishes on the basis of a system of values. Most societies are class societies in which the upper classes are said to need to live on a much higher material standard than the lowest, in order to avoid degradation (a major social calamity!). But are these wishes or needs?

The so-called basic needs, those necessary for survival, are only made fixed magnitudes through verbal magic. And 'survival' is a term of little use if restricted to mere 'not dying.' Remember the final words of Chief Seattle on the great change the White Man would bring to the land: 'the end of living and the beginning of survival.'

The transition from the discussion of such ethically basic norms to more political norms may be formulated in many ways. Here is one.

(1) The requirement of minimum conditions of Self-realization should have priority before others.

(2) This requirement implies that of minimum satisfaction of biological, environmental, and social needs.

(3) Under present conditions many individuals and collectivities have unsatisfied biological, environmental, and social needs, whereas others live in abundance.

(4) To the extent that it is objectively possible, resources now used for keeping some at a considerably higher level than the minimum should be relocated so as to maximally and permanently reduce the number of those living at or below the minimum level.

One can say that the derivation of basic norms in Ecosophy T splits in two different directions. The last level we have outlined presents the norms and hypotheses of the local community, a characteristic ideal of many utopian systems. Now we are ready to follow an argument towards politics to justify the norms and hypotheses against exploitation, as developed through debates with the Marxists in Norway in the early 1970s.

H11: Self-realization requires realization of all potentials.
H12: Exploitation reduces or eliminates potentials.
  N9: No exploitation!
H13: Subjection reduces potentials.
N10: No subjection!
N11: All have equal rights to Self-realization!
H14: Class societies deny equal rights to Self-realization.
N12: No class societies!
H15: Self-determination favors Self-realization.
N13: Self-determination!

Comments:

The above formulations are put forth mainly to show that the fundamental norms of Self-realization do not collide with norms of increasing the reign of justice on Earth. On the contrary, the class differences inside societies and between nations are clearly differences in conditions of Self-realization. Exploitation may be defined in terms of semi-permanent or permanent reducing of the possibilities of some groups in favor of others. Furthermore, calculations showing differences in the use of energy and other resources support an ecological approach in the fight against exploitation in class societies. The value of the model consists partly in the derivation of a general political attitude or posture without the use of certain terms such as 'communism,' 'socialism,' 'private enterprise,' and 'democracy' which elicit more or less automatic positive or negative reactions.

(g) The overview of Ecosophy T in diagram form

All these formulations (N1 to N13, H1 to H15) contain key terms from social, political, and life philosophy. They do no more than

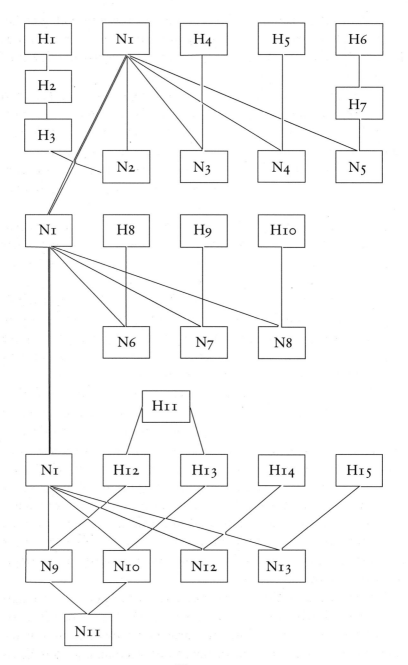

Figure 2

suggest how the systematization of more precise norms and hypotheses could be related. Nothing more is pretended. Figure 2 illustrates in schematic form the logical derivations of these first four levels of Ecosophy T.

The slogan-like character of these formulations still deserves to be used in environmental debate, but there is today clear awareness of their limitations. In many cases local communities have fought sane ecological policies and invited disastrous development. And rather strong central authorities are required to implement the national and international policies recommended by the *World Conservation Strategy* (1980).

The above way of depicting logical derivations within an ecosophy has been greeted with an interest that is in part misplaced: what can be obtained by the normative systematization of the illustrated kind is very modest, only a better survey of a few interconnections within a gigantic unsurveyable whole. After all, a total view, a philosophy centered on human-nature relations, touches upon so many complex questions that the explicit formulations can only comprise a small part, especially when the formulations attain a decent level of preciseness and vague slogans are shunned.

A person using partial systematizations of Ecosophy T will normally find reasons (through increasing experience of life) to modify the formulations. The T_0-formulations can only act as guidelines, and modifications will normally consist of changing accepted precisations of these formulations. To find a formulation on the T_0 level false, mistaken, or invalid is a rather strange thing. It implies that one finds every plausible interpretation false, mistaken, or invalid.

Rejections in terms of falsehood or invalidity will occur on the levels of more precise formulations—formulations of which the set of plausible interpretations makes up a proper subset of the original formulation at the T_0 level.

The logical derivation diagram of a fragment of Ecosophy T cannot easily be broadened to accommodate a whole ontology. Field or gestalt thinking goes against the status of the distinction human-environment as it is met in most environmental thinking. Here Ecosophy T joins contemporary non-Cartesian philosophical trends.

Looking towards the next century, the shock-waves of ecology will reach so far and penetrate so much that the flora of eco-terminology may seem redundant. The ecosophies will, I suppose, be absorbed in the general traditions of philosophy of nature (*Naturphilosophie*). In all this I may be wrong, but the signs of change through internalization are already present.

5

Platform Principles
of the Deep Ecology Movement

Arne Naess and George Sessions

In April 1984, during the advent of spring and John Muir's birthday, George Sessions and Arne Naess summarized fifteen years of thinking on the principles of the deep ecology movement while camping in Death Valley, California. In this great and special place, they articulated these principles in a literal, somewhat neutral way, hoping that they would be understood and accepted by persons coming from different philosophical and religious positions.

Readers are encouraged to elaborate their own versions of deep ecology, clarify key concepts, and think through the consequences of acting from these principles.

Basic Principles

1. The well-being and flourishing of human and nonhuman Life on Earth have value in themselves (synonyms: intrinsic value, inherent value). These values are independent of the usefulness of the nonhuman world for human purposes.

2. Richness and diversity of life forms contribute to the realization of these values and are also values in themselves.

3. Humans have no right to reduce this richness and diversity except to satisfy vital needs.

4. The flourishing of human life and cultures is compatible with a substantial decrease of the human population. The flourishing of nonhuman life requires such a decrease.

5. Present human interference with the nonhuman world is excessive, and the situation is rapidly worsening.

6. Policies must therefore be changed. These policies affect basic economic, technological, and ideological structures. The resulting state of affairs will be deeply different from the present.

7. The ideological change is mainly that of appreciating *life quality* (dwelling in situations of inherent value) rather than adhering to an increasingly higher standard of living. There will be a profound awareness of the difference between big and great.

8. Those who subscribe to the foregoing points have an obligation directly or indirectly to try to implement the necessary changes.

Naess and Sessions Provide Comments on the Basic Principles:

Re (1). This formulation refers to the biosphere, or more accurately, to the ecosphere as a whole. This includes individuals, species, populations, habitat, as well as human and nonhuman cultures. From our current knowledge of all-pervasive intimate relationships, this implies a fundamental deep concern and respect. Ecological processes of the planet should, on the whole, remain intact. "The world environment should remain 'natural'" (Gary Snyder).

The term "life" is used here in a more comprehensive nontechnical way to refer also to what biologists classify as "nonliving": rivers (watersheds), landscapes, ecosystems. For supporters of deep ecology, slogans such as "Let the river live" illustrate this broader usage so common in most cultures.

Inherent value as used in (1) is common in deep ecology literature. ("The presence of inherent value in a natural object is independent of any awareness, interest, or appreciation of it by a conscious being.")

Re (2). More technically, this is a formulation concerning diversity and complexity. From an ecological standpoint, complexity and symbiosis are conditions for maximizing diversity. So-called simple, lower, or primitive species of plants and animals contribute essentially to the richness and diversity of life. They have value in themselves and are not merely steps toward the so-called higher or rational life forms. The second principle presupposes that life itself, as a process over evolutionary time, implies an increase of diversity and richness. The refusal to acknowledge that some life forms have greater or lesser intrinsic value than others (see points 1 and 2) runs counter to the formulations of some ecological philosophers and New Age writers. Complexity, as referred to here, is different from complication. Urban life may be more complicated than life in a natural setting without being more complex in the sense of multifaceted quality.

Re (3). The term "vital need" is left deliberately vague to allow for considerable latitude in judgment. Differences in climate and related factors, together with differences in the structures of societies as they now exist, need to be considered (for some Eskimos, snowmobiles are necessary today to satisfy vital needs).

People in the materially richest countries cannot be expected to reduce their excessive interference with the nonhuman world to a moderate level overnight. The stabilization and reduction of the human population will take time. Interim strategies need to be developed. But this in no way excuses the present complacency—the extreme seriousness of our current situation must first be realized. But the longer we wait the more drastic will be the measures needed. Until deep changes are made, substantial decreases in richness and diversity are liable to occur: the rate of extinction of species will be ten to one hundred times greater than in any other period of Earth history.

Re (4). The United Nations Fund for Population Activities in its State of World Population Report (1984) said that high human population growth rates (over 2.0 percent per annum) in many developing countries "were diminishing the quality of life for many millions of people." During the decade 1974–1984, the world population grew by nearly 800 million—more than the size of India. "And we will be adding about one Bangladesh (population 93 million) per annum between now and the year 2000."

The report noted that "The growth rate of the human population has declined for the first time in human history. But at the same time, the number of people being added to the human population is bigger than at any time in history because the population base is larger."

Most of the nations in the developing world (including India and China) have as their official government policy the goal of reducing the rate of human population increase, but there are debates over the types of measures to take (contraception, abortion, etc.) consistent with human rights and feasibility.

The report concludes that if all governments set specific population targets as public policy to help alleviate poverty and advance the quality of life, the current situation could be improved.

As many ecologists have pointed out, it is also absolutely crucial to curb population growth in the so-called developed (i.e., overdeveloped) industrial societies. Given the tremendous rate of consumption and waste production of individuals in these societies, they represent a much greater threat and impact on the biosphere per capita than individuals in Second and Third World countries.

Re (5). This formulation is mild. For a realistic assessment of the situation, see the unabbreviated version of the I.U.C.N.'s *World Conservation Strategy*. There are other works to be highly recommended, such as Gerald Barney's *Global 2000* Report to the President of the United States.

The slogan of "noninterference" does not imply that humans should not modify some ecosystems as do other species. Humans have modified the Earth and will probably continue to do so. At issue is the nature and extent of such interference.

The fight to preserve and extend areas of wilderness or near-wilderness should continue and should focus on the general ecological functions of these areas (one such function: large wilderness areas are required in the biosphere to allow for continued evolutionary speciation of animals and plants). Most present designated wilderness areas and game preserves are not large enough to allow for such speciation.

Re (6). Economic growth as conceived and implemented today by the industrial states is incompatible with (1)–(5). There is only a faint resemblance between ideal sustainable forms of economic growth and present policies of the industrial societies. And "sustainable" still means "sustainable in relation to humans."

Present ideology tends to value things because they are scarce and because they have a commodity value. There is prestige in vast consumption and waste (to mention only several relevant factors). Whereas "self-determination," "local community," and "think globally, act locally," will remain key terms in the ecology of human societies, nevertheless the implementation of deep changes requires increasingly global action—action across borders.

Governments in Third World countries (with the exception of Costa Rica and a few others) are uninterested in deep ecological issues. When the governments of industrial societies try to promote ecological measures through Third World governments, practically nothing is accomplished (e.g., with problems of desertification). Given this situation, support for global action through nongovernmental international organizations becomes increasingly important. Many of these organizations are able to act globally "from grassroots to grassroots," thus avoiding negative governmental interference.

Cultural diversity today requires advanced technology, that is, techniques that advance the basic goals of each culture. So-called soft, intermediate, and alternative technologies are steps in this direction.

Re (7). Some economists criticize the term "quality of life" because it is supposed to be vague. But on closer inspection, what they consider to be vague is actually the non-quantitative nature of the term. One cannot quantify adequately what is important for the quality of life as discussed here, and there is no need to do so.

Re (8). There is ample room for different opinions about priorities: what should be done first, what next? What is most urgent? What is clearly necessary as opposed to what is highly desirable but not absolutely pressing?

6

Arne Naess and the
Union of Theory and Practice

George Sessions

While giving a series of Schumacher Lectures on the deep ecology movement in England in January of 1992 Professor Arne Naess celebrated his 80th birthday. Naess' influence upon me over the last 15 years has been immense: as a friend, philosophical colleague, mentor, and fellow supporter of the long-range deep ecology movement. In many ways, our backgrounds before we met were similar: strong interests in philosophy, the philosophy of science, and ecology; rock climbing and a love of mountains and wild nature; and an awareness of the need to work out an ecological philosophy as a basis for social change and a long-range solution to the environmental crisis. Through a long intellectual and emotional journey, I finally came to realize in the late '70s that Arne Naess had arrived at the main outlines of such a position many years earlier. Gary Snyder independently worked out a similar ecological position in the '60s, and has been an inspiration and mentor to me as well (see Jon Halper, *Gary Snyder: Dimensions of a Life*, Sierra Club Books, 1991).

Arne and I also independently shared the conviction that Spinoza, more than any of the other major Western philosophers, provided a good model and inspiration for a contemporary ecological philosophy. Arne's interest in Spinoza began as a young boy of 17 and in conjunction with early experience in the mountains; my interest developed as a result of surveying the history of Western philosophy around 1969–70. I was particularly impressed by the American philosopher Charles Frankel's comparison, in 1955, of the philosophical variation of the pragmatists, James and Dewey, with that of George Santayana (who was heavily influenced by Spinoza). Frankel pointed out that:

> For James and Dewey, the future is open. Nature and the past are just raw materials for us to do with as we wish. . . . Santayana rejected pragmatism, was irritated by it for this reason. He felt that there is, in the human environment, a certain resisting structure, a certain permanent constitution of things to which we must all ultimately bow down . . . [Santayana] seems more of a philosopher to me; a man more concerned with looking at the world "in the aspect of eternity," to use Spinoza's phrase . . ." [he] represented a kind of wisdom from which the pragmatists might have learned." (As quoted in my "Spinoza and Jeffers on Man in Nature," *Inquiry,* Oslo, 1977)
>
> The views of Spinoza and Santayana express not only a respect for Nature, but an awareness of limits on human activity in Nature.

During my reading of this period, I was also impressed by the observations of Stuart Hampshire (in his 1951 book on Spinoza) which led me to believe that, in my search, Spinoza was the philosopher I was looking for:

> In Descartes and in Leibniz, one is still in various ways given the impression of a universe in which human beings on this earth are the privileged center around whom everything is arranged, almost, as it were, for their benefit; whatever their professed doctrine, almost everyone still implicitly thought in terms of a man-centered universe . . . [Spinoza] had this inhuman vision of human beings as not especially significant or distinguished parts of an infinite system, which seems in itself vastly more worthy of respect and attention than any of our transitory interests and adventures. (As quoted in my "Anthropocentrism and the Environmental Crisis," *Humboldt Journal of Social Relations,* 1974)

One should not misinterpret the Spinozistic assessment of humanity's place in the cosmos and on the Earth as misanthropic, but rather as a major reappraisal of Western culture's approach to the question: "What is humanity's place in Nature?" For example, Professor Naess sees humans as "very special beings!" The problem, in his view, is that we humans "underestimate our potentialities" both as individuals and as a species. Our abilities to understand and identify with Life on Earth suggest a role primarily *as appreciators* of the biotic exuberance and evolutionary processes on Earth, rather than as conquerors, dominators, manipulators, or controllers and "business managers" of the Earth's evolutionary processes.

But from this mainly abstract metaphysical approach to developing an ecophilosophy (as a result of my classical and analytical training in academic philosophy), except for the anti-anthropocentrism, I still remained essentially within the secular mechanist modernist paradigm. The Spinozistic perspective did lead me to see, at this point, that an adequate ecophilosophy must go beyond the development of merely a new "environmental ethics" to a radically different worldview (a "paradigm change" if you will). This change of paradigm to an organic/spiritual worldview began to occur for me from the middle to late '70s with the reading and digesting of Roszak's *Where the Wasteland Ends*, Capra's *Tao of Physics*, and the writings of Alan Watts and Gary Snyder on Zen Buddhism, together with my own experiences of deep connectedness with wild nature. Jacob Needleman's *A Sense of the Cosmos* had a profound psychological influence on me. I had to read it many times to begin to understand it, although Needleman, in my opinion, fails to make the bridge to an identification with wild nature. I put together and taught a college course beginning about 1975 called "Rationality, Mysticism, and Ecology," using the books by Roszak, Watts, Snyder, and Needleman.

In 1973 I had heard about Arne Naess from Joseph Meeker. Several years earlier, I had heard from a UC Berkeley student in Yosemite about a Norwegian philosopher who was teaching a course that strongly emphasized ecology and Nature, and using poetry from Robinson Jeffers. About 1975, I was wandering around an academic hall at the University of the Pacific, named for my late grandfather, George Colliver, who founded the religious studies department there, when I by chance came across an announcement on a bulletin board calling for graduate students to study Spinoza and ecology with Arne Naess in Norway under the New Philosophy of Nature program. My interest in Naess perked up considerably at this point and we began to correspond and exchange papers. When I finally met Naess at UC Santa Cruz in 1978, I remember going out to a beach to the north, and Arne Naess taking up with a pack of dogs, running gleefully back and forth in the surf playing tug-of-war with long strands of kelp.

I was now open to understanding Spinoza from a more organic/psychological/spiritual perspective. I was helped in this by the forceful and humorous personality of the late UC Santa Barbara philosopher, Paul Wienpahl, and by his Zen Buddhist organic approach to Spinoza in his book *The Radical Spinoza* (1979). A fuller appreciation of the significance of Spinoza's spiritual/psychological analysis of the free human individual (and the crucial distinction between active and passive behavior) was due to Arne Naess. He has pointed out that:

> Part Five of [Spinoza's] *Ethics* represents, as far as I can understand, Middle East wisdom par excellence.... The free human being is a wise human being permanently and with increasing momentum on the road to still higher levels of freedom. The supremely free person shows perfect equanimity, forceful, rich and deep affects, and is active in a great variety of ways corresponding to the many "parts of the body," and all of them bound up with increasing understanding—and certainly including social and political acts.... This image of the sage has in common with (a certain variety of) Mahayana Buddhism the idea that the higher the level of freedom reached by an individual, the more difficult it gets to increase the level without increasing that of all other beings, human and nonhuman.... It again rests on identification with all beings. (Naess, "Through Spinoza to Mahayana Buddhism, or Through Mahayana Buddhism to Spinoza?" in Wetlesen, *Spinoza's Philosophy of Man*, University of Oslo Press, 1978)

Part Five of Spinoza's *Ethics* (although the earlier metaphysics and epistemology of the book provide the supporting worldview) is actually a description and guide for the spiritual/intellectual character development of a certain type of mature person. Professor

Naess exhibits this type of character and personal development to a very high degree, spiced with a delightful and ever present sense of humor. And it is this type of "perennial philosophy" spiritual development (to use Aldous Huxley's phrase), found throughout most cultures down through history and pre-history, including primal peoples, which sets this tradition off, along with other significant differences, from the many frivolous and questionable "spiritual" activities of the so-called New Age movement. This spiritual tradition, and sense of what constitutes a mature human, also provides a sharp contrast to the secular cultural relativism of modern industrial societies, which seems to justify much of the thoughtless and callous exploitation and abuse of both humans and nonhumans, thus leading to the highly immature destructive economic orientation and consumerism of modern life.

Arne Naess' Spinozist and Gandhian perspective on human maturity and spiritual development must also be reconciled with, and tempered by, his love and respect for *diversity* (human, cultural, biological, etc.)—for example, Naess truly loves and finds fascinating the tiniest insect—and his philosophic scepticism concerning attaining ultimate human knowledge. Hence, the significance and importance placed on the endless Socratic deep questioning process in "deep" ecological inquiry. When we went to Death Valley in 1984 to work on the deep ecology movement platform, I came to appreciate his genuine openmindedness and insistence upon a diversity of philosophic and religious positions.

In the "apron diagram" depicting the different derivational levels of the deep ecology platform, it is recognized that differing ultimate premises, and philosophical and religious commitments (and cultures) at Level 1 of the diagram are in many ways incompatible; this widely diverse situation, however, is to be desired, not deplored. Supporters of the deep ecology movement will, however, tend to agree on the main points of the *platform,* or on some similar alternative formulation [for the apron diagram and the platform, see earlier chapters in this Anthology]. Arne Naess points out that very few supporters of the deep ecology movement are professional philosophers who would work out, or subscribe to, the intricacies of a logically derived Level 1 philosophical position, such as Arne's Ecosophy T, but everyone, using the deep questioning process, should attempt to make their overall worldview and ultimate commitments clear, at least to themselves, thus becoming aware of how they are related to specific actions and practical decisions in environmental and everyday life situations.

Warwick Fox, in his recent book (*Toward a Transpersonal Ecology*), has suggested that Arne's philosophical approach to the deep ecology movement should be thought of in terms of the "wide identification" thesis, of a psychological widening identification with all species—a "transpersonal ecology." But as Naess points out, Fox's thesis is one among many kinds of alternate ecosophies, a Level 1 set of commitments that not every supporter of the deep ecology movement does, or would hold. The deep ecology movement is much broader, is partly characterized by the deep questioning process and the platform, and should not be identified with any particular Level 1 ecosophy.

Recently, Naess has pointed out that his philosophical work can be divided into four periods or phases. The first period, up through about 1940, concentrated on the philosophy of science. The second period, from about 1940 through 1953, consisted of work in empirical semantics. A short third period concentrated on anti-dogmatism and the revival of "the largely forgotten classic Greek Pyrrhonic scepticism." The fourth period began about 1968, partly at the urging of his students, when his interests shifted to ecological philosophy. At this point, he tells us, he began to live his philosophy and function in part more as a "minor prophet" than as a strictly academic philosopher. In this respect, his life seems to parallel the life of a philosopher like Bertrand Russell, who began as a strict academician in mathematics, philosophy of mathematics, epistemology, and metaphysics; but, late in life, became more and more engrossed in social and practical issues such as peace, disarmament, and avoiding nuclear war. But unlike Russell, Arne has developed, and continues to develop, a unique philosophically sophisticated ecophilosophy, and his own personal Level 1 Ecosophy T—a sort of unique summation of his earlier work in mathematical logic, philosophy of science, empirical semantics, philosophical/cultural anthropology, Pyrrhonic scepticism, Gestalt perception and epistemology, and his important original studies and publications in Spinoza and Gandhi (a blending of the wisdom of East and West)—as a philosophic basis for his role as contemporary world ecophilosopher, promoter of the deep ecology movement, and "minor prophet."

From Gandhi (as well as Spinoza), Naess reinforced his intuitions about the equal "right" of all species to flourish on the planet (ecocentric egalitarianism) together with the Eastern doctrine of *ahimsa*—the avoidance of causing unnecessary harm and suffering. Naess likes to tell the story about Gandhi refusing to let people in the ashram kill poisonous snakes or other creatures. In lectures, he shuffles his feet across the floor to show how the residents of the ashram avoided stepping on the snakes to avoid being bitten. And from Gandhi came theories and strategies for nonviolent action and interaction with people, in word and deed. Arne is deeply committed to, and has written extensively on, positive supportive nonviolent communication between people, another tie-in with his work in empirical semantics, and his life-long interest in language. Naess, in his incredibly gentle and sophisticated way, constantly reminds those of us who write in philosophical ecology not to respond in kind to cruel or unfair attacks on the deep ecology movement, and to strongly condemn ecologically destructive *acts*, such as the clearcutting of the last ancient forests in North America, but not the *persons* who perform these acts. Many of us find this a difficult ideal to live up to, a sort of exercise in Bertrand Russell's solution to Meinong's "golden mountain" paradox, but Naess is, of course, right about this.

Finally, I would like to say something about Arne Naess' personal lifestyle, a lifestyle which goes hand-in-hand with his role as "ecological prophet"—a role he shares with Gary Snyder (with his deep ecological exemplary bioregional living), and with Dave Brower (also 80 years old, the senior statesman of the American environmental movement, and referred to as "Muir reincarnate"). At a symposium in May of 1983 at York University in Canada (sponsored by John Livingston, Neil Evernden, and the Faculty of Environmental Studies), Naess presented a short paper on "Deep Ecology and

Lifestyle." He talked about tendencies among the supporters of the deep ecology movement toward (1) using simple means; (2) anti-consumerism; (3) appreciation of ethnic and cultural differences; (4) efforts to satisfy vital needs rather than desires; (5) going for depth and richness of experience rather than intensity; (6) attempts to live in nature and promote community rather than society; (7) appreciating all life forms; (8) efforts to protect local ecosystems; (9) protecting wild species in conflicts with domestic animals; (10) acting nonviolently, etc., to which he later added a tendency toward vegetarianism.

Perhaps the most striking of the tendencies he listed was (11) "concern about the situation of the third and fourth world and the attempt to avoid a standard of living too much different from and higher than the needy. Global solidarity of lifestyles." which was closely followed by (12) "Appreciation of lifestyles which are universalizable, which are not blatantly impossible to sustain without injustice towards fellow humans or other species." On these criteria, highly consumerist lifestyles in industrialist countries are totally unjustifiable in relation to the living conditions of the poor of the world, quite apart from the immense negative impact they are causing in the destruction of wild ecosystems and biodiversity throughout the world. This follows "ethically" from the degree of *identification* we experience with our fellow humans, and with other species and ecosystems (with our degree of maturity). And Naess outdoes the most ardent "social justice" proponents in industrial countries by proposing lifestyles not "too much different from and higher than the needy." Industrial societies ... may resist the idea that consumption levels in these countries will have to be drastically reduced, while living standards in Third World countries must dramatically improve, together with human population stabilization and reduction among both the rich and the poor countries, it is now clear that there will be no solution to the environmental crisis without such a move.

Arne Naess himself lives this lifestyle. The Naess family is one of the most influential, and highly regarded families in Norway. As chairman of the philosophy department at the University of Oslo for about 40 years, Professor Naess was instrumental in helping give the Norwegian public school system a decidedly philosophic cast; most students have read the history of philosophy he authored, and studied his book *Empirical Semantics*. Naess has been surrounded by immense financial wealth most of his adult life; his two older brothers, who live in New York and the Bahamas, are shipping magnates, as is his nephew, Arne Naess, Jr., whom Arne raised as his son. Arne, Jr., took up Arne's love for mountain climbing, leading a Norwegian expedition to the summit of Mt. Everest before marrying the singer, Dianna Ross. At the age of 70, Arne accompanied the expedition to base camp at 20,000 ft.

In the great philosophic/religious spiritual traditions of both East and West, from Buddha and Socrates to Ludwig Wittgenstein (the famous early 20th-century Viennese philosopher who gave away a large family inheritance to lead an austere philosophic life), Naess likewise leads a Spartan existence, giving away half his pension each year to worthy causes, such as reroofing a schoolhouse in Nepal. Eating Arne's diet is a humbling experience, as I discovered on our 1984 trip to the desert, and on other occasions. We stopped for breakfast one morning, walked out into the mesquite and built a tiny fire where we toasted one slice of bread apiece (no butter or anything). His favorite main course is a "stew" consisting of cut up potatoes and carrots with no seasoning. It certainly leads to robust health and stamina on his part; I have a vision of him walking off into the desert, backpack on, carrying a gallon of water in each hand. In the spring of 1991, while staying with Michael Soulé, the conservation biologist, Naess biked from the UC Santa Cruz campus down to the ocean and back. He commented that he must be slipping; he had a much harder time getting back up the hill than he had the year before. He made the run again, and then realized he had the bike in the wrong gear. He seemed considerably relieved.

Naess loves the deserts of the South West United States and Baja California, with their incredible diversity of wildflowers in the spring. He likes to walk two or three miles from the road, build a little fire to brew some tea, ideally by some small cliffs where he can rock climb, read philosophy, and work on and rewrite his many papers.

There are pictures of him sitting in a mountain tent, high on a Himalayan peak, reading Spinoza's *Ethics*. I remember him once visiting me in the Sierra foothills, whereupon he immediately built a tiny fire out of twigs, sat down, and watched the wisp of smoke drift down toward the canyons. He once said to me, "George, you are so lucky to still have such a richness of wild areas here in the United States!" He recently visited Antarctica, camping out in the snow in his little tent for days.

When he was in his 20s, he built a little cabin high on his favorite mountain in Arctic Norway (the highest hut in Norway). His father died when he was very young, and this mountain became a kind of father-substitute for him. It is said that this hut has had up to seven roofs on it at once, held down with cables, and that, in a severe storm, several roofs can fly off. Naess has spent up to six months at a time living up there. It is well stocked with books. The stories about the severe conditions there are legion and the ice sometimes never completely melts inside the hut. I think Arne Naess is happiest when he is there; doing philosophy, climbing rocks, skiing, floating on the little pond when it isn't iced over, enjoying the company of the friends and loved ones who go with him, and living his austere lifestyle.

Many of Naess' philosophical papers (both straight academic philosophy and ecophilosophy) are unpublished or are scattered around in obscure journals. By and large, his tremendous philosophic achievement is unknown to the academic philosophic community around the world, although the importance of his work is well recognized in Norway. His work on the deep ecology movement, however, is now well known and has brought him into the center of ecophilosophical and environmental circles throughout the world. Just as Russell, Wittgenstein, and Einstein were seen as the leading Western intellects of the first part of the 20th century, I am confident that when Naess' overall philosophic achievement is synthesized and published, he will be recognized as a leading philosopher of the latter half of the twentieth century. For his work in developing an ecological philosophy and paradigm, and by articulating and helping to launch the long-range deep ecology movement, he may well be recognized as one of the most important philosophers of the 20th century.

Section II

Elaboration

7

Re-Inhabitation

Gary Snyder

I came here by a path, a line, of people that somehow worked their way from the Atlantic seaboard westward over a hundred and fifty years. One Grandfather ended up in the Territory of Washington, and homesteaded in Kitsap County. My Mother's side was railroad people down in Texas, and before that they'd worked the silver mines in Leadville, Colorado. My Grandfather, being a homesteader, and my father a native of the state of Washington, put our family relatively early in the Northwest. Yet we weren't early enough. An elderly Salish Indian gentleman came by our farm once every few months in a model T truck, selling smoked salmon. "Who is he?" "He's an Indian" my parents said.

Looking at all the different trees and plants that made up my second-growth Douglas fir forest plus cowpasture childhood universe, I realized that my parents were short on a certain kind of knowledge. They could say "That's a Doug Fir, that's a cedar, that's bracken fern ..." But I perceived a subtlety and complexity in those woods that went far beyond a few names.

As a child I spoke with the old Salishan man a few times over the years he made these stops—then, suddenly, he never came back. I sensed what he represented, what he knew, and what it meant to me: he knew better than anyone else I had ever met, *where I was*. I had no notion of a white American or European heritage providing an identity; I defined myself by relation to the place. Soon I also understood that "English language" is an identity—and later, via the hearsay of books, received the full cultural and historical view—but never forgot, or left, that first ground: the "where" of our "who are we?"

There are many people on the planet, now, who are not "inhabitants." Far from their home villages; removed from ancestral territories; moved into town from the farm; went to pan gold in California—work on the Pipeline—work for Bechtel in Iran. Actual inhabitants—peasants, paisanos, paysan, peoples of the land, have been sniffed at, laughed at, and overtaxed for centuries by the urban-based ruling elites. The intellectuals haven't the least notion of what kind of sophisticated, attentive, creative intelligence it takes to "grow food." Virtually all the plants in the gardens and the trees in the orchards, the sheep, cows and goats in the pastures were domesticated in the Neolithic; before "civilization." The differing regions of the world have long had—each—their own precise subsistence pattern developed over millennia by people who had settled in there and learned what particular kinds of plants the ground would "say" at that spot.

Humankind also clearly wanders. Four million years ago those smaller proto-humans were moving in and out of the edges of forest and grassland in Africa; fairly warm; open enough to run in. At some point moving on, catching fire, sewing clothes, swinging around the arctic, setting out on amazing sea voyages. A skull found in Santa Barbara has been dated at 50,000 years. So it may be that during the middle and late Pleistocene, large fauna hunting era, a fairly nomadic grassland-and-tundra hunting life was established, with lots of mobility across northern Eurasia in particular. With the decline of the ice age—and here's where we are—most of the big game hunters went out of business. There was possibly a population drop in Eurasia and the Americas, as the old techniques no longer worked.

Countless local ecosystem habitation styles emerged. People developed specific ways to be in each of those niches: plant knowledge, boats, dogs, traps, nets, fishing—the smaller animals, and smaller tools. From steep jungle slopes of Southwest China to coral atolls to barren arctic deserts—*a spirit of what it was to be there* evolved, that spoke of a direct sense of relation to the "land"—which really means, the totality of the local bio-region system, from cirrus clouds to leaf-mold.

So, inhabitory peoples sometimes say "this piece of land is sacred"—or "all the land is sacred." This is an attitude that draws on awareness of the mystery of life and death; of taking life to live; of giving life back—not only to your own children, but to the life of the whole land.

Abbé Breuil, the French prehistorian who worked extensively in the caves of southern France, has pointed out that the animal murals in those 20,000-year-old caves describe fertility as well as hunting— the birth of little bison and cow calves. They show a tender and accurate observation of the qualities and personalities of different creatures; implying a sense of the mutuality of life and death in the food chain; and what I take to be a sense of a sacramental quality in that relationship.

Inhabitation does not mean "not travelling." The term does not of itself define the size of a territory. The size is determined by the bio-region type. The bison hunters of the great plains are as surely in a "territory" as the Indians of northern California, though the latter may have seldom ventured farther than 30 miles from where they were born. Whether a vast grassland, or a brushy mountain, the Peoples knew their geography. Any member of a hunting society could project from his visualization any spot in the surrounding landscape, and tell you what was there; how to get there. "That's where you'd get some cattails." The bushmen of the Kalahari desert could locate a buried ostrich egg full of emergency water in the midst of a sandy waste—walk right up and dig it out, "I put this here three years ago, just in case."

Ray Dasmann has useful terms to make these distinctions: "ecosystem-based cultures" and "biosphere cultures." By that Ray means societies whose life and economies are centered in terms of natural regions and watersheds, as against those who discovered—seven or eight thousand years ago in a few corners of the globe—that it was "profitable" to spill over into another drainage, another watershed, another peoples' territory, and steal away its resources, natural or human. Thus the Roman Empire would strip whole provinces for the benefit of the capital, and villa-owning Roman aristocrats would have huge slave-operated farms in the south using giant wheeled plows. Southern Italy never recovered. We know the term "imperialism"—Dasmann's "biosphere cultures" adds to that, helps us realize that biological exploitation is a critical part of it too—the species made extinct. The clearcut forests.

All that wealth and power pouring into a few centers had bizarre results. Philosophies and religions based on fascination with society, hierarchy, manipulation, and the "absolute." A great edifice called "the state" and the symbols of central power—in China what they used to call "the true dragon"; in the West, as Mumford says, symbolized perhaps by that bronze age fort called the Pentagon. No wonder Lévi-Strauss says that civilization has been in a long decline since the Neolithic.

So here in the twentieth century we find occidentals and orientals studying each other's Wisdom, and a few people on both sides studying what came before both—before they forked off. A book like *Black Elk Speaks*, which would probably have had zero readership in 1900, is perceived now as speaking of certain things that nothing in the Judaeo-Christian tradition, and almost nothing in the Hindu-Buddhist tradition, deals with. All the great civilized world religions remain primarily human centered. That next step is excluded, or forgotten—"well, what do you say to Magpie? What do you say to Rattlesnake when you meet him?" What do we learn from Wren, and Hummingbird, and Pine Pollen, and how. Learn what? Specifics: how to spend a life facing the current; or what it is to perpetually die young; or how to be huge and calm and eat *anything* (Bear). But also, that we are many selves looking at each other, through the same eye.

The reason many of us want to make this step is simple, and is explained in terms of the 40,000-year looping back that we seem to be involved in. Sometime in the last ten years [1967–1977] the best brains of the Occident discovered to their amazement that we live in an Environment. This discovery has been forced on us by the realization that we are approaching the limits of something. Stewart Brand said that the photograph of the Earth (taken from outer space by a satellite) that shows the whole blue orb with spirals and whorls of cloud, was a great landmark for human consciousness. We see that it has a shape, and it has limits. We are back again, now, in the position of our Mesolithic forebears—working off the coasts of southern Britain, or the shores of Lake Chad or the swamps of southeast China, learning how to live by the sun and the green at that spot. We once more know that we live in a system that is enclosed in a certain way; that has its own kinds of limits, and that we are interdependent with it.

The ethics or morality of this is far more subtle than merely being nice to squirrels. The biological-ecological sciences have been laying out (implicitly) a spiritual dimension. We must find our way to seeing the mineral cycles, the water cycles, air cycles, nutrient cycles, as sacramental—and we must incorporate that insight into our own personal spiritual quest and integrate it with all the wisdom teachings we have received from the nearer past. The expression of it IS simple: gratitude to it all, taking responsibility for your own acts; keeping contact with the sources of the energy that flow into your own life (i.e., dirt, water, flesh).

Another question is raised: Is not the purpose of all this living and studying the achievement of self-knowledge, self-realization? How does knowledge of place help us know the Self? The answer, simply put, is that we are all composite beings, not only physically but intellectually, whose sole individual identifying feature is a particular form or structure changing constantly in time. There is no "self" to be found in that, and yet oddly enough, there is. Part of you is out there waiting to come into you, and another part of you is behind you, and the "just this" of the ever-present moment holds all the transitory little selves in its mirror. The Avatamsaka ("Flower Wreath") jeweled-net-interpenetration-ecological-systems-emptiness-consciousness tells us, no self-realization without the Whole Self, and the whole self is the whole thing.

Thus, knowing who and where are intimately linked. There are no limits to the possibilities of the study of who and where, if you want to go "beyond limits"—and so, even in a world of biological limits, there is plenty of open mind-space to go out into.

Summing Up

In Wendell Berry's essay "The Unsettling of America" he points out that the way the economic system works now, you're penalized if you try to stay in one spot and do anything well. It's not just that the integrity of Native American land is threatened, or National Forests and Parks; it's *all* land that's under the gun, and any person or group of people who tries to stay there and do some one thing well, long enough, to be able to say, "I really love and know this place," stands to be penalized. The economics of it works so that anyone who jumps at the chance for quick profit is rewarded—doing proper agriculture means *not* to jump at the most profitable chance—proper forest management or game management means doing things with the far future in mind—and the future is unable to pay us for it right now. Doing things right means living as though your grandchildren would also be alive, in this land, carrying on the work we're doing right now, with deepening delight.

I saw old farmers in Kentucky last spring who belong in another century. They are inhabitants; they see the world they know crumbling and evaporating before them in the face of a different logic that declares, "everything you know, and do, and the way you do it, means nothing to us." How much more the pain, and loss of elegant cultural skills, on the part of non-white fourth-world primitive remnant cultures—who may know the special properties of a certain plant, or how to communicate with Dolphins, skills the industrial world might never regain. Not that special, intriguing knowledges are the real point: it's the sense of the magic system; the capacity to hear the song of Gaia *at that spot*, that's lost.

Re-inhabitory refers to the tiny number of persons who come out of the industrial societies (having collected or squandered the fruits of 8000 years of civilization) and then start to turn back to the land, to place. This comes for some with the rational and scientific realization of inter-connectedness, and planetary limits. But the actual demands of a life committed to a place, and living somewhat by the sunshine green plant energy that is concentrating in that spot, are so physically and intellectually intense, that it is a moral and spiritual choice as well. "*Mankind* has a rendezvous with destiny in Outer Space." Some say. We are already travelling in space. This is the galaxy, right here. The wisdom and skill of those who studied the universe first hand, by direct knowledge and experience, for millennia, both inside and outside themselves, is what we might call the Old Ways. Those who envision a possible future planet on which we continue that study, and where we live by the Green and the Sun, have no choice but to bring whatever science, imagination, strength, and political finesse they have to the support of the inhabitory people—natives and peasants of the world. Entering such paths, we begin to learn a little of the Old Ways, which are outside of history, and forever new.

[Based on a talk given at the "Re-inhabitation Conference" at North San Juan School, held under the auspices of the California Council on the Humanities, August 1976.]

8

Shifting Paradigms:
From Technocrat to Planetary Person

Alan Drengson

In this essay I examine the interconnections between two paradigms of technology, nature, and social life, and their associated environmental impacts. The dominant technocratic philosophy which now guides policy and technological power is mechanistic. It conceptualizes nature as a resource to be controlled fully for human ends and it threatens drastically to alter the integrity of the planet's ecosystems. In contrast, the organic, person-planetary paradigm conceptualizes intrinsic value in all beings. Deep ecology [movement principles] give priority to community and ecosystem integrity and help to guide the design and applications of technology according to principles which follow from ecological understanding. I will describe this shift in paradigms and how it affects our perceptions, values, and actions.

The Problems

We hardly need to remind ourselves of the extent and seriousness of contemporary environmental problems. The episode at Three Mile Island, the recent discoveries of the extensive problems caused by irresponsible disposal of toxic wastes, such as at Love Canal in New York, the accelerating oil and energy prices, the threat of nuclear war, these are only a few examples in the news. Since the first Earth Day, in 1970, increased public attention has been focused on the three major areas: overpopulation, pollution, and resource depletion.

Numerous reform measures have been attempted, sometimes with limited success, but on the whole the index of environmental quality has declined over the last decades. We do not seem to have made significant progress, and many areas where progress once seemed possible have more recently come under renewed pressure. Decisions are made on political and narrow economic grounds, rather than on principles that are environmentally sound, and economically advantageous in the *long* term.

Various analyses have been offered about what should be done to safeguard the environment and human welfare from the hazards of modern technologies. These technologies have given us much in the way of comfort and enjoyment, and they have made possible the development of human skills on a scale never before possible. But they are also perceived by many as Frankenstein monsters loose among us. In films, novels, art, poetry, and even within the scientific establishment there are many voices of disquiet, fearing a human society controlled and imprisoned by its own technological creations. There is much discussion of alienation and anomie, the loss of community, the impersonal character of machine-like bureaucracies. Some writers, for example, Robert Heilbroner in *An Inquiry into the Human Prospect*, see virtually no chance that we can avoid major disasters, given our current direction.[1] Heilbroner suggests that to control these problems we will need powerful central governments with tight controls that are incompatible with democracy. It seems as if our technology has an alien, monster-like character. As a culture we are of two minds about it. On the one hand, we welcome its advantages, but on the other, we often refuse to acknowledge its shortcomings. During the last ten years we have certainly become more aware of the dual nature of the uses of technology.

Many analyses of the ecological crisis have emphasized an inseparability of ourselves from our environment. Arne Naess, a Norwegian philosopher, distinguishes between those who see the problems in isolated ways compatible with mild reform and those who see the problems holistically as requiring a deep change in our form of life. Only the latter, according to Naess, will put our relationships with our ecosystems on a sound, co-evolving basis.[2] The essential features of the shallow ecology movement are its mild reformist character and its anthropocentric bias that the nonhuman world has only instrumental value. The shallow ecology movement is essentially oriented towards the health and well-being of the peoples of the advanced industrial nations. The deep ecology movement, however, recognizes the need for a fundamental shift to new paradigms of human-environmental relationships in terms of fields and processes. The deep ecology movement emphasizes, in principle, biospheric egalitarianism and the intrinsic value of all life.[3] It aims for the creation of systems that are diverse, symbiotic, and compatible with natural systems, and for the establishment of an anticlass attitude which is consistent with biospheric egalitarianism. Deep ecological knowledge forces us to recognize that ecosystems are in some cases so complex that they will probably never be completely understood by humans. Thus, supporters of the deep ecology movement humbly recognize human ignorance, and the need for cautious development of technology. They seek to avoid the fragmentation and complication of human life that results from too great an emphasis on technological control.

In this essay, I explore the possibility of a shift in paradigms such that we can be freed from outlooks that lead to serious environmental problems, and can then move to those that promote environmentally sound and sustainable human societies. I employ the Kuhnian notion of paradigms and try to make clear the almost unconscious role they play in conditioning our thought.[4] We do tend to become captives of our own metaphors and models, just as we do of their associated techniques. A creative, flexible approach requires that we be able to shift our perspectives in ways more appropriate to our problems. We must become aware not only of our constraints, but also of our larger possibilities.

Humans organize and orient their lives in terms of various ideals, models, symbols, and metaphors. A major function of myth is to weave knowledge, aspirations, and skills together in an intersubjective realm of image and symbol that blends art and science. Mythic symbols can store and convey vast amounts of meaningful information in concise form. This makes it possible for a person to assimilate the collective experiences of his/her culture. In a loose sense dominant paradigms are forms of mythic understanding. The technocratic paradigm has a powerful mechanistic image from which a large number of other subordinant paradigms develop, such as methods of analysis, forms of technique, and the like. However, when pursued to its logical extremes, it seriously undermines the vigor of other aspects of human life, such as the intersubjective, which are important. In my view this paradigm has reached its limits, as evidenced by the stresses between it and the work of contemporary science, by its apparent negative cultural effects, and by its consequences on the environment. Our culture seems to be going through a major transition to a postindustrial culture. Its exact shape cannot be determined beforehand, but it is possible to discern some of the main components of this shift.

Evolution requires that information be processed and encoded in various ways in an organism. We inherit a repertoire from the past, but as we live in the world we also have to improvise. In human life much learning is accumulated and passed on to the young by cultural means. In some respects this makes greater flexibility possible; however, if taken too absolutely this becomes a form of mental conditioning which can result in lost flexibility. A saving antidote here is a healthy dose of Socratic ignorance, humbly recognizing our limitations and the relativistic character of our theories about the world.[5] Socratic ignorance forces us to be open and to reclaim the transparent character of inquiry pursued with full awareness. Thus, one aim in contrasting paradigms is to free our minds so that we can look at the world afresh. If we view paradigms as art (or literary) forms, we can then better appreciate the need to avoid conceptual rigidity. In what follows I describe in broad outline the major features of the two dominant paradigms, the technocratic and the person-planetary.

On Paradigms

Thomas Kuhn points out that Western science did not develop historically as the result of simple accretion. It has been characterized by periods of deep change, followed by times of consolidation and linear growth, in which there is a subsequent elaboration upon the theories and models created during revolutionary periods. The result of creative periods is a set of modes of explanation, methods, laws, theories, and instruments which govern scientific orthodoxy during periods of nonrevolutionary change. A paradigm is a constellation of models which defines, exemplifies, and illustrates the ideals and procedures of normal science during nonrevolutionary times. However, as all paradigms have limitations, the labors of normal science itself eventually lead to a breakdown of these paradigms by exposing their limitations. This in turn forces the creation of new paradigms.

Particular paradigms exist in every field of specialization. They are part of what students learn when they study to be physicists, biologists, economists, psychologists, or philosophers. When the paradigms of a particular discipline are more productive in a given era than others, these tend to be deployed by other disciplines as well. Thus, certain paradigms of the physical sciences have come to be widely applied, not only to all natural sciences, but also to the social sciences and, in limited ways, to the humanities. In these fields too we find attempts at precision and objectivity, experimental methods, quantification and analysis, methods modeled after the paradigms of the physical sciences, especially as these came to be defined by mechanistic conceptions of the world. Older, more holistic ways of thinking were supplanted by methods of analysis and experimentation that aimed primarily at prediction and control. This positivistic orientation has in turn led to a fragmentation of our knowledge of nature, humans, societies. However, the data and information developed by "hard" studies has, at the same time, tended to undercut the paradigms which have guided their application. Thus, a paradigm shift is developing in some of the more theoretically advanced studies.

When a major cultural activity, such as science, undergoes a paradigm shift, human perceptions are changed, since we interpret the world in terms of the paradigms that are in dominant use. The notion of paradigm shifts then, as we employ it here, is not restricted to the development of science, but is extended to cultures as a whole.

My thesis as set forth below is that our culture is undergoing, and is in need of, a major paradigm shift. Further, the emerging paradigm, if in support of the deep ecology movement, will be more appropriate to the unity and the interrelatedness of the Earth, with its limitations and its delicately balanced ecosystems. Such an orientation will stimulate more fulfilling personal development as well. The emerging ecological paradigm I call the person-planetary; the waning, older paradigm the technocratic.[6]

The spirit of inquiry and creativity comes alive during such periods of paradigm shift. We are more open to novelty and the multifarious complexity of the world. We are able to see new possibilities. Our lives take on added dimensions of significance. To be sure, the openness of such times risks both conceptual confusion and a conceptless desert in which orientation can be difficult. In our current situation, one force driving our civilization toward fundamental change is the confusion created by lack of understanding of our own technological forces. Part of this confusion arises from the fact that science, art, and philosophy have become separated. The older models no longer work as explanations for this process of technological development. Nor do they work as they ought for direction of policy. We are overwhelmed by large amounts of information and highly specialized learning. The older paradigm no longer seems compatible with, nor able coherently to connect our vast knowledge. Even our experience of the world is fragmented, rather than united. We need new unifying insights capable of enabling us better to understand human and ecosystem processes.

In the discussion which follows we recognize that within each of the two dominant outlooks there are subordinate, supporting paradigms, but we will refer to the collective paradigms of each outlook in the singular as the technocratic or the person-planetary paradigm. We also note that major paradigms have sufficient power to survive cultural change. For example, organic paradigms have characterized philosophies in past cultures. This means that the current spiral of development is in some ways open to an infusion of elements of older wisdom. The emerging organic paradigms will differ from older organic paradigms, for detailed knowledge of ecosystem processes, cells, matter-energy, planetary processes and the like inform the emerging paradigm. Nonetheless, features of earlier organic paradigms can be incorporated so as to give continuity over a longer time span. Moreover, the waning paradigm is rarely totally rejected but aspects of it are incorporated in the new paradigm. So the paradigm of the technocratic outlook will have useful but limited applications.

The Technocratic Paradigm

Descartes' conception of methods of inquiry, added to the Baconian idea of putting nature on the rack of questions, helped lay the foundations of modern science and philosophy. Until recently, Cartesian philosophy of technique has dominated modern disciplines.[7] Philosophy has typically extended Cartesian analysis by formalizing all modes of inquiry so that they resemble mathematics (understood sometimes as logic), or, if it has rejected some specific techniques, it has retained the Cartesian philosophy of technique. The fundamental idea is that proper application of right technique will in time yield solutions to any problem. This emphasizes the uniformity not only of method but also of the character of the problems. We aim to reduce phenomena to their component parts, and we explain all wholes by reference to these parts and to their external, measurable relationships. So conceived, technique sets the program for technology, applied in uniform ways to production. Labor is so divided that each component need perform but one technique. It is easy to apply this idea to human life in general, including thought processes. Thus ideas too have their simplest "parts." New ideas result from recombinations, just as new compounds are formed from the rearrangement of atomic elements. The whole universe begins to look like a complex machine.

A masterful technique has a power that is almost irresistible. It seems to simplify our problems and our lives. Unfortunately, the power of technique can become a substitute for understanding. To be sure, for a time concentration on technique can expedite our learning. However, all techniques have limitations. Moreover, complete mastery in some cases can mean a transcendence of technique, and this carries possibilities of a creativity which develops new techniques. It seems a human compulsion to explore fully each new technique, to elaborate it, to consider all of its possible variations. This ensures that the limitations of any technique must eventually be realized. Thus their complete elaboration often leads to their overthrow. Here we have an example of dialectical interaction. Every idea implies, i.e., logically connects with, its negation, its opposite. The extreme elaboration of a practice often generates an opposing reaction. But opposites interpenetrate. We tend to move from one extreme to the other as we overreact to situations judged in polar terms. Or we have a tendency to do more of what we are doing, when we dread the opposite result. This often has the consequence of bringing that very result to pass. Thus, attempting to arrest social change can often help precipitate drastic change.

Powerful techniques solve many problems but eventually often generate problems of their own. The techniques of modern science and technology are prone to create higher level problems, when their limits are reached; technology tends to become itself the prime concern of modern industrial culture. Thus we now perceive a need for control of technology by experts, either directly or through elected or corporate officials whose decisions will be guided by technicians. The amounts of capital involved in technology, the numbers of people, the skill levels required, the increasing risks dictated by the required economics of large-scale investment, long lead time, etc., all press us toward a government by an expert elite. Perhaps the actual politicians and decision makers will not necessarily be experts, except in modern management techniques, but the net cultural effect of such technological development will result from the values of the experts, which underlie their decisions. In time this cultural bias leans toward technocracy.

Technocracy here refers to the systematic application of technology to all levels of human activity, including governmental and economic policies which have growth as their central aim. Such growth in the contemporary West is often promoted by means of policies which favor complex, high technologies. The scale involved in applying new technologies dictates a need for government and corporate planning; thus, only specialists can write policy. The aim becomes the control of life by means of management techniques that govern the application of the hardware and processes integral to technology. Science is narrowed to its less theoretic activities and the principal emphasis is upon prediction, control, and applied science. The sciences so stressed are thought to be value-free. The aim is to reduce all phenomena to those features which can be quantified, controlled, and observed directly with the instruments produced by technology.

By these means we *objectify* persons and nonhuman nature. Intersubjective experience cannot be "captured" or characterized in these terms. Thus it becomes subjective in the pejorative sense, unreal, unimportant, irrelevant. Ironically this thoroughgoing objectifying ultimately undercuts its own reason for being, since it denies access to the whole of human experience and finds ends beyond its means to evaluate. Technology can only be a tool, a means, despite the fact that good design can create products of intrinsic aesthetic merit. Technocracy must then rest on ends beyond its own capacity to understand or justify. Carried to its logical end it seeks to turn the world into a controlled artifact. Nature is only a resource to be processed. This process in turn becomes self-perpetuating and self-justifying; and in time it must also bring human social activities under technological control. This in turn involves behavioral technology and social engineering. Humans must now be "designed" to fit the technological mold, since they are fallible in their normal human form and might interrupt the technological process.

While the technocratic paradigm partly defines what science should be, it does not preclude that science might consider values. But even if this were to be done, the result would attempt to reduce them to nonvalues. In any case, the practical social result, since technological priorities determine the flow of research money, is that those areas of human endeavor which promote technocracy will thrive under technocratic paradigms, whereas those that promote deeper values will languish. Think of the large amount of resources devoted to the hard sciences and engineering, compared with the small amount devoted to aesthetics, art, and the humanities.

Descartes held that creatures with souls have significant intrinsic value. But once Descartes' conception of soul and God are rejected by natural and humanist philosophers, technocracy is free to emphasize methods and means, now impoverished in ends. The emphasis upon method and technique, upon reductionism, upon explaining all natural phenomena in mechanistic terms, the quantification of as much in the natural world as possible, these are emphasized by those who keep only the materialistic half of Cartesian dualism. This materialism is of the classical atomistic variety. These elements, then, more than any others, defined the positivistic shape that the philosophy of science assumed. As was indicated earlier, these elements were closely tied to the development of modern industrial manufacturing. This explains in part the emphasis on increased production and consumption, for this is the means to measure efficiency, even though these measures do not reflect total costs.

The technocratic mindset strives to create the perfect machine process at all levels of society. The machine metaphors for the body, for nature, for the solar system, and for social systems have been illuminating in limited ways. During the last three hundred years these models have penetrated Western consciousness as more and more of its energies have been directed to the creation of modern, industrial, machine-based technologies and economies. The sheer intensity of this effort, coupled with the logic of these technologies and their anthropocentric values, often seems destined to literalize these metaphors. Thus the Earth comes to be seen as a machine, devoid of consciousness but for humans, and even in humans the methods of empiricist science pass consciousness by, or attempt to technologize it. Technocratic philosophy even makes it difficult to distinguish consciousness from machine "intelligence." All of this is done, supposedly, for greater human interests, as none of the other planetary inhabitants have any value in their own right. Thus the technocratic "machine" drives to manage all aspects of natural, industrial, and social processes by means of centralization, substituting where possible machines for humans, rules and laws for morality, social system for community, monoculture for diversity, and so on. This drive is found in capitalist and socialist nations alike, for the mechanistic paradigm is essentially placeless, global, transpolitical, transideological, and is closely connected with modern industrial technology and its specialized disciplines.

Just as nature comes to be treated only as a resource, so persons are evaluated on the basis of their functions, rather than in terms of their intrinsic worth. Production of things becomes more important than persons and communities. Jobs, progress, the glory of the state, the facility of computer systems, efficiency, these are cited as justification for disrupting persons, families, and communities. The technocratic state emphasizes wealth, power, and the capacity to influence others. Although lip service may be paid to helping others, the system of rewards and sanctions ensures that those who choose service will be under rewarded, while those who strive for power will be rewarded in material wealth and prestige. The forms of organization that arise are all corporate entities, whether business, university, government, or military. These all converge in the technocratic state. Diversity is discouraged, for such a state tends to become a monolithic monoculture. Voluntary associations and public interest groups may form in reaction, but their influence will be small, their resources meager. Friendship and genuine community are difficult within corporate structures, since corporations demand loyalties which often conflict with these values.

In summary, the technocratic paradigm encourages the development of centralization, of capital-intensive and labor-poor industry. It strives to apply technology to all human life and creates uniformity in product and culture. It fragments human life and finally lacks any values to sustain it. As it becomes more centralized and complicated, it becomes more vulnerable to the "Titanic effect."[8] The concept of persons is impoverished. Nature is understood, not as living subject, but as object and machine. The intrinsic value of intersubjective community life, as exemplified in art, fiction, storytelling, folklore, myth, poetry, and drama, lacks serious recognition within the technocratic paradigm. In this discussion we have emphasized the negative aspects of the technocratic paradigm in order to sharpen the contrast we will next draw with the person-planetary paradigm. All these inadequacies the ecological paradigm is meant to correct, because it develops from an understanding of community, and because it is more adequately based on contemporary insights into the human and natural world.

The Person-Planetary Paradigm

The essence of Socratic wisdom is found through the clear perception of our own ignorance. Thus armed, we are freed from the boundaries that our self-assertive interests draw between ourselves and others. When we arrogantly attempt to measure all things by human measures, when we think we have at last controlled life's uncertainties and eliminated its mysteries, our view seems absolute. We allow it to dictate how we should live, even when this view is itself a major source of our problems. We find it difficult to connect our philosophy with the quality of our experience, with our actions and their consequences. In such a case, Socrates would direct us to the way of dialogue and the cure of dialectic. We are limited not by an aware ignorance, but by our "knowledge" that the world is as we think it "must" be. In contrast, Socratic ignorance has the virtue of freeing the intelligence to follow the inquiry wherever it might lead. In this case, our quest is an enlarged paradigm for the creation of social processes which will be in harmony with a broader perception of ecosystem health.

Each of us knows in our bones that the world is not a machine. Nor are our bodies mechanisms. Computers are neither intelligent nor conscious. Poetry is as significant as mathematics. The most valuable things in life cannot be measured or quantified. Both friendship and genuine community are necessary for the development of whole human persons. Nature is not some alien monster that opposes us, that we must conquer. Tigers and wolves are not just killing-machines; even these fierce hunters are capable of tenderness and affection. Nature is not completely predictable. Humans, and at least some animals, are aware that the other is sometimes a subject. Most of us do not hate the natural world. We are deeply moved by its beauty, awed by its majesty and power. We do not wantonly destroy or pollute it. Each of us knows that we are not separate from it.

Yet there is pollution and destruction. Our collective actions cause serious consequences. These sometimes seem overwhelming, as if produced by an impersonal technology over which we have no control. Part of our inherited culture takes the form of the technocratic paradigm, but there are other elements, such as those mentioned above. We will now develop an organic person-planetary paradigm to see how it might better integrate these insights with our contemporary knowledge of ecology, humans, and community. We begin with a reflection on social philosophy since 1600.

Prior to the scientific revolution, the dominant Western philosophy of humans and nature was Christian. The world was seen as created by an act of God's love. Humans were created in His image. The world was theirs in trust from Him. Their role was to have dominion over it, but also to care for His creation. Nature was thus sacralized. Humans would destroy it only at the peril of eternal damnation. However, when the world came to be conceptualized as a machine, when modern methods of control began to be applied, there was tension between this Christian outlook and the scientific views of humanism. Essentially, humanism took over many Christian values, but it especially emphasized priority of human dominion over the world. In harmony with modern science humanism emphasized human ability to understand the world, to demystify it, to desacralize it, and to control it. As both God and soul were left out of humanist philosophy, there eventually arose the technocratic doctrine that we have discussed, which stressed individualistic separatism, utilitarianism, mechanism, and anthropocentrism.[9] Nature becomes a secular object, resources to be used to satisfy human cravings. The only locus of value is seen as pure subjective preference, or as a calculable "greatest good for the greatest number of humans."

Kant observed, in contrast to Hobbes, that "humans are not only self-assertive, self-oriented and antisocial, but they also desire sociability, not simply to be admired personally, but also because they are social beings."[10] Community life is intrinsically valuable to us. It is the dialectical interplay between these conflicting drives that creates society. Kant echoes the Baconian warning that nature is to be commanded only by obeying her. Civilization cannot exist outside the realm of nature, for natural laws provide the constraints within which society must exist. Freedom in its highest expression involves acting on principles. In the social matrix these constraints and freedoms are balanced out. In community one realizes one's worth as a person acting in concert with others in a kingdom of ends.

If Hobbes emphasizes our separateness, Kant emphasizes community. Technocratic philosophy, with Hobbes, regards each of us as separate parts which get their significance only by being related through the state by means of laws externally imposed on us. Our connection to the world and to other persons, then, is through externalities which define the range of possible relationships, and which also deny us the significance of relationships that unite us with other subjects in a meaningful community context. Person-planetary philosophy, with Kant, regards community as primary. Through its paradigm, observer and observed are united in reciprocal processes of interresponsiveness. The boundaries of community extend to include the other beings of our home places. We affect and are affected by this broader community of life. Our societies are living processes within it.

The person-planetary paradigm attempts to locate the constraints on human activities in the principles of ecology and the reality of particular ecosystems. These reveal that ecosystems are more like organisms than machines. The interrelationships between organisms within an ecosystem are not completely specifiable, unlike the case of a machine. There are elements of variability and unpredictability. Various elements of balance are so complexly interrelated that they intersect and double back on themselves, form pyramids in symbiotic complexities that magnify and also minimize effects. If one does apply a machine model to an ecosystem, this can be done only for abstracted, large "components," and even then it is a kind of Rube Goldberg machine, *qua* machine. Ecosystems are both entropic and antientropic.

The dominant organic paradigm of the person-planetary stresses the interrelatedness of the biosphere. The world must be thought of as intersecting fields of processes, rather than as separate individuals. We cannot isolate our actions from the rest of society, nor from the rest of the ecosystem. Polluting the water in the stream that runs through my yard can pollute all water in the drainage. The ground water polluted by radioactive wastes can pollute the river, the ocean, the biosphere. Unlike a machine, the organism is a complexly inter-related whole of processes, with both internal and external princi-ples of organization. The ecosystem is like a living body.

We have some understanding of the vast complexity of the human body as a result of centuries of accumulated empirical study. Under-standing does not necessarily mean power to control, nor does power to control necessarily mean one understands. One weakness of the machine model is that it gives us the illusion of understanding when we "explode" the parts and see them in display. But something is fully understood only when, after analysis, we can view the subject as whole once more. Some move toward synthesis is necessary. As Lewis Thomas points out, we cannot understand the cell in isola-tion, but only in relation to higher levels of integration.[11] The body is a community of cells. In a larger human community persons have a range of freedom, but their mutual ends finally harmonize and this makes communion possible. If each is an isolated Hobbesian per-son with unlimited drives of self-assertion, then there is no alterna-tive to external control. If we cannot be internally self-regulating persons within a context of consciously shared values, then we can only be regulated from without. This is the conclusion drawn within the technocratic paradigm and this logic drives it toward complete control. Further, this control will not first be applied to the larger collective social processes, but rather to persons. Since persons are the social atoms, the aim will be to bring them into conformity with the ends of the technocratic state. Ultimately this control has no jus-tification other than its own power to maximize the welfare of iso-lated individuals, who are now denied freedom and intrinsic worth by this very control.

The ancient Chinese sage Lao Tzu was one of the world's early philosophers of ecology. In the *Tao Te Ching* he observed that all things are equal in the great natural order. The trouble begins when we start to separate ourselves from this order. We do this first by passing judgments which attempt to elevate ourselves over other beings. The human impulse to manage the world is an expression of the judgment that we know best how the natural world should run. Ironically, we find every day that we do not know enough, and probably never will know enough, to prevent the unfortunate consequences of attempting to manage too much. We did not know that DDT would ultimately reach the Arctic, nor that it would pollute even human milk. We did not know that aerosols would threaten the atmosphere. We did not know that nuclear power would pose the risk of the *n*th country in nuclear arms. The list goes on endlessly. Failure of our systems is blamed on human error, not on our lack of knowledge, not on the limits of our power, not on the arrogance of anthropocentrism, not on our basic philosophy. It is always the fallible human part that is said to be at fault, never the whole approach.

From the perspective of the deep ecology movement, managerial attempts to control the natural world create difficulties to the extent to which our design ignores the values of other beings and natural ecosystems. There are not only human values at stake, but also the values of all other organisms. Biospheric egalitarianism and the principles of ecological interconnection help us to realize that no large-scale impacts on ecosystems will be without their effects on human life. The greater the effect observed on other life forms, probably the greater will be the effect on us. Since social processes are interrelated as well, ecological principles must be introduced at the inception, not at the conclusion of design.

To refuse to recognize the intrinsic worth of other beings, to fail to appreciate the subtle ways in which natural processes work, and to seek centralized control is finally to be saddled with the ultimate responsibility that once was thought to be God's. Humanism, as anthropocentrism, joined with the technocratic paradigm, must finally assume the overwhelming responsibility for running everything. All nature must be managed for human ends, and even these ends must be managed. Ultimately, to value humans alone is to leave us without value, for then we are unable to find value in the world; value becomes purely subjective. The deep ecology movement would reclaim value by placing it once more at the center of life, and by broadening our conception of human experience consonant with this.

With respect to the intrinsic worth of each being, no philosopher in our tradition has expressed this better than A.N. Whitehead:

Everything has some value for itself, for others, and for the whole. This characterizes the meaning of actuality. By reason of this character, constituting reality, the conception of morals arises. We have no right to deface the value experience which is the very essence of the universe. Existence, in its own nature, is the upholding of value intensity. Also, no unit can separate itself from the others, and from the whole. And yet each unit exists in its own right. It upholds value intensity for itself, and this involves sharing value intensity with the universe. Everything that in any sense exists has two sides, namely, its individual self and its significantion in the universe. Also, either of these aspects is a factor in the other.[12]

Compare Whitehead's remarks with those of Justice William O. Douglas in a dissenting opinion on Mineral King, a California wilderness area threatened by development:

The river, for example, is the living symbol of all the life it sustains or nourishes—fish, aquatic insects, water ouzels, otter, fisher, deer, elk, bear, and all other animals, including man, who are dependent on it or who enjoy it for its sight, its sound, or its life. The river as plaintiff speaks for the ecological unit of life that is part of it. Those people who have a meaningful relation to that body of water—whether it be a fisherman, a canoeist, a zoologist, or a logger—must be able to speak for the values which the river represents and which are threatened with destruction.[13]

Both Whitehead and Douglas, then, recognize that humans can perceive the intrinsic value in natural processes, animals, other beings. In Douglas's case a variety of relationships with the river are mentioned, but each meaningful perspective on the values of the river implies that these values are not human alone.

One of the most difficult matters in environmental issues has to do with representation of interests. Courts have often taken a narrow view of "interests." In the case of Mineral King the Court denied standing to the Sierra Club on the grounds that the club members would not be injured, and did not have an interest in the proposed development of Mineral King. The Sierra Club redrafted its brief to name members aggrieved by the proposed resort. What the Court in effect said was that only humans, or the fictional persons of ships and corporations, can be given standing. The majority view is that environmental disputes must be settled, through the legal channels, by means of resolving conflicts of interests. For example, in the case of a forest, the forest as such cannot be aggrieved. Conservationists have interests, but these must be *their* interests, not the values within a natural process, independent of human interests. Douglas's dissenting opinion suggests that other living beings, and processes like the river, have their own values which should be recognized by the Court. Those who are qualified by their meaningful relationships with the subject in question should be allowed to represent these values.

The technocratic model treats all interests as human interests. Aesthetic values, species values, recreational, habitat, and other values of a forest are to be quantified in terms of common dollar values, and these are to be weighed against the monetary values to be gained by logging, or by some other economic use. We tend to bridle at the suggestion that all human values can be meaningfully quantified, let alone given a dollar value. It is much more difficult to place a dollar value on natural processes. Moreover, if we do not recognize value in the natural world, if we are rigidly committed to the fact/value distinction, then we are confronted with the problem of trying to find any value at all. If there are only human values, but no values in the world apart from our interests, do human values then have any objective meaning? Would this not force us ultimately to consider all differences in value perception as conflicts of interest, in which the most powerful interest will always win?

Whitehead avoids this difficulty by a metaphysics that takes value as part of the meaning of actuality. Value experience, he says, is the very essence of the universe. This value that each being has for itself is also shared by others. So each exists for itself, but also exists for the other. It is a value in itself, and a value for others. It has both intrinsic and instrumental value. Both Douglas and Whitehead, then, would transcend narrow anthropocentrism. Both recognize the interpenetration of things with one another. The values spoken of imply diversity, one of the features of most ecosystems. Intrinsic value in each being denies the appropriateness of centralized control. The value of the river is for many others, but it has a value in itself. These intersecting value relationships within the context of ecosystems promotes flexible stability. The features of diversity, flexibility, adaptation, symbiosis, accommodation, interconnectedness, all suggest a process of design for human technologies that deeply understands these processes, an understanding that is enriched by the perception of both intrinsic and instrumental values. This understanding is not the result of "objective" study alone, but requires that we also approach the subject with a respect for its way that is free of one-sided judgments, for these obstruct a deeper appreciation of these diverse forms of life.

How then do we come to this understanding? Consider the words of Rolling Thunder:

> Too many people don't know that when they harm the earth they harm themselves, nor do they realize that when they harm themselves they harm the earth. . . . It's not very easy for you people to understand these things because understanding is not knowing the kind of facts that your books and teachers talk about. I can tell you that understanding begins with love and respect. It begins with respect for the Great Spirit, and the Great Spirit is the life that is in all things—all creatures and plants and even the rocks and the minerals. All things—and I mean *all* things—have their own will and their own way and their own purpose; this is what is to be respected. Such respect is not a feeling or an attitude only. It's a way of life. Such respect means that we never stop realizing, and never neglect to carry out our obligations to ourselves and our environment.[14]

What is implied by Rolling Thunder's view is that we cannot understand the ways of other beings, so long as we approach them only *via* interests, economic or personal. We must also be able to approach them on their own terms, through love and respect, as a way of being, as a way of acting in relation to them. Any other approach separates us from other beings, narrows our perspective, encroaches upon our poetic and aesthetic responses by drawing boundaries through judgments. Rolling Thunder sees that we and the environment are not separable. Conflict with the world is therefore a conflict that begins within us. Conflicts of interest narrowly defined are then only conflicts about how we shall *use* the world for our benefit alone. They do not reflect any deep appreciation of things in themselves. Rolling Thunder tells us it is possible to appreciate the value of other beings, but only through respect.[15]

It might be said that we do not know what other natural beings value. What do deer and bear want? But the point is that this is not for us to judge. They have their way of life. This we should respect. We share our lives with others. Our communities include other beings. To disrespect them is to impoverish our own lives, to deprive ourselves of their values. It is, moreover, to pave the way to destructive exploitation, for once we refuse to recognize any values other than human values, there is no way to recover these lost values short of a deep readjustment in our way of thinking, feeling, and acting toward the world. Our various dichotomies, as between human and non-human, reason and emotion, fact and value, hinder experiencing the world in ways fully receptive to other lives. Rolling Thunder suggests that the way of love and respect will open us up once more, unite us with the world, and enable us then to experience the values of other beings.

Our traditional moral philosophies cannot resolve these environmental issues and conflicts.[16] These philosophies do not recognize values in natural things and they are individualistic in their conceptions of rights and duties. Our collective actions are really beyond their reach. The problems will not be resolved by denying the value of individuals, so that they might be completely subjugated to the interests of the collective. Nor will they be solved by denying responsibility for our collective activities. Each individual has values in itself, for itself. The dandelion that grows in the meadow has its own value, but it also shares this with others in its biotic community. Individuals do not exist in isolation, but in relationship. We only begin to understand this in a deep way when we begin to appreciate the value of their way. Such a respect clearly implies a different orientation toward the natural world and also toward human life. One paradox of Western democracies is that we uphold individual values even when the "persons" in question are corporate and their actions have collective impact; yet, we often attempt to control individual persons by laws which prosecute victimless "crime." We often punish persons for harm they might do to themselves, and thus we do not respect their freedom and dignity. Yet we have no effective way of curbing corporate activities that impinge upon the rights of each person by polluting the air (s)he breathes or corrupting the water (s)he drinks. Clearly these imbalances indicate that our system is in a period of great stress.

The person-planetary paradigm helps us to understand the mutual interpenetration of living communities and their ecosystems. The interconnections are evident on several levels of integration. From the energy-flows through ecosystems, the pyramid structure of living organisms in terms of population, the networks of food chains, the relationships of neighborhood and territory, the interpenetration of reciprocal awareness in the active responses of sentient beings, the consciously shared ends and purposes in human communities, all of these represent levels of integrated organization within the total ecosphere. In human communities persons live within the rich cultural and ideational matrix where their feelings are interwoven with those of others. We tend to live in this culturally conditioned consciousness without noticing the role played by dominant metaphors in shaping our reactions to others. Moreover, we are often unaware of the extent to which we lose our capacity for wholehearted response through habit, and through being captured by these dominant models. The technocratic paradigm begins its control of nature with a control of our minds, which affects how we see the world and what we look for.

The person-planetary paradigm shift enables us to look at the world through the eyes of ecological processes and relationships. Afterwards, we begin to see the need for and the way of designing our processes so that they are compatible with the principles of ecology and consistent with a respect for living beings in ecosystem contexts. We can better tell when a process complements an ecosystem rather than seriously altering it. The eye for proper "fit" is, in one respect, the eye uncolored by any theory or judgment, but nonetheless informed by deep experience in the broad sense. In such eyes art and science meet. In emphasizing intrinsic worth and egalitarianism, the person-planetary paradigm helps us to see the possibility of designing collective activities that from their very conception are ecospherically sound, capable of coevolving with other life forms. Thus the organic paradigm leads us from individual actions to an appreciation for persons and beings in communal contexts. These contexts lead to the processes that interconnect communities globally. From the person, human and nonhuman, and the community, we are led finally to see the planet as a whole. (The story of the whole Earth . . .) A view of the whole Earth suggests that it too can be illuminatingly seen as an organism (biography). The recently proposed Gaia hypothesis represents a rebirth of the ancient wisdom that the Earth is a living mother to us all. Seeing the planet as a living being reinforces our understanding of the interconnectedness of biospheric processes. There is, in a real sense, a symmetry between our bodies and the body of the planet.

Let us liken the world to a living symphonic poem. It is not a symphony in which a central score fully determines the music yet to come, nor does each performer have its role rigidly fixed. Each has its own characteristic voice in this jazz symphony. We do not know where it is going, since it is improvised as we go along. Each voice fits itself to the music made by all the others, but each has a chance to play its own tunes. (It is inter-responsive.) To learn to fit ourselves into this ongoing harmony requires attention and agility. For purposes of design, philosophical and otherwise, we need conceptual agility. Organic art requires a sensitive awareness which is receptively responsive to the music as it is played, immediately aware of the symphony as a whole and of its many voices. If we are attuned to our world in this way, control is no longer an issue. That is to say, we realize we do not have to try to manage everything else. We can then learn to approach things on a smaller, more decentralized scale, based on subtle understanding. We can learn to ride with the natural processes, rather than against them. We will, in short, trust nature because we will trust ourselves appropriately to improvise as the need arises, rather than trying to follow some rigid plan that we impose on everything else, and even on ourselves.

The shift to an organic paradigm means loosening the hold of the mechanical approach, but that has become maladaptive and insecure in any case. Pure "reason" and "objectivity" devoid of sensitive awareness, aesthetic response, insight, intuition, or caring attention prevent leaving the "security" of this narrowed experience. This narrowness is itself a source of insecurity. If we narrow our experience in this way, then we tend to live incompletely. This almost ensures that we will be dissatisfied and insecure. Life becomes a tangle of unfinished problems that we plan to solve in the future by more tightly controlling the world. However, by doing our best as whole persons we can realize that the natural world will support us indefinitely, if we but adjust our activities to it. We can also realize how we can open to a centered awareness that is free of conceptual constraints. This openness in turn can lead to a realization of our essential being and inherent worth, as it is in the world as a whole.[18] We do not need to distinguish ourselves to realize our value. We have nothing to prove. Our deep ecological paradigm sees human life at home once more in a harmonious world. We are not isolated and alone. We are citizens of the world, members of the larger communities of life.

Summary

The person-planetary paradigm stresses: internal principles of order and the importance of homeostasis and balanced development; context and place; symbiosis and mutual interrelationships, decentralization, diversity and unity, spontaneity and order, freedom in community; intrinsic value in being itself, biospheric egalitarianism, human experience as value-laden; creative, ecologically compatible design of human activities; collective responsibility and the unique value of individuals, personal knowing, intersubjective experience and diverse consciousness; organisms as wholes which interact with other organisms in spheres of interpenetration; the planet as a whole as a living organism; persons as creative, open, dynamic, developmental, and as coevolving within larger communities. The technocratic paradigm stresses: atomistic analysis; reductionism, mechanism; context-free abstractions; anthropocentrism; plurality and isolation; determinism and laws; eternal principles of order; manipulation and centralized control; repetitive and predictable patterns of action; interchangeable parts; value-free experience, objective, abstract, disinterested observation; value in nature as only instrumental; persons as mechanical, closed, in need of control, capable only of linear growth.

If our Western Institutions shift to the person-planetary paradigm as our primary orientation, we will have some chance of creating postindustrial processes that will not destroy ecosystem resilience and vitality. However, if we continue in the technocratic mode there is less chance of this since capital-intensive, large-scale programs will undermine all attempts to add on environmental constraints at the end of the design process. At that point "cost-benefit" and similar analyses can be used to justify expediency and "efficiency." Should these fail, then "national emergency" or similar appeals will always suffice.

To a certain extent our paradigm selection is a creative affair, but by this we do not mean to suggest that it is entirely arbitrary. Many paradigms are so limited that they cannot be generally applied. Furthermore, my main argument in this essay has been that the principles of ecology and the current state of advanced scientific knowledge point toward emerging metaphysical conceptions of the world that are more in keeping with an ancient outlook and organic paradigms, than with the machine paradigm of the technocratic outlook.[19] Philosophy has a significant contribution to make in helping to develop ecocentric paradigms that are ecophilosophical in the broad sense, and which represent a creative synthesis of current knowledge. What I have offered here is only a preliminary sketch.

Notes

1. Robert L. Heilbroner, *An Inquiry into the Human Prospect* (New York: W.W. Norton, 1974).

2. Arne Naess, "The Shallow and the Deep, Long-Range Ecology Movement. A Summary," *Inquiry* 16 (1973) 95–100.

3. As Naess explains, "The 'in principle' clause is inserted because any realistic praxis necessitates some killing, exploitation, and suppression" (Ibid., p. 95). This qualification is made with the simple recognition that we cannot live without affecting the world to some degree. In the discussion that follows, whenever biospheric egalitarianism is discussed this qualification is assumed.

4. Thomas S. Kuhn, *The Structure of Scientific Revolutions* (Chicago: University of Chicago Press, 1970).

5. On Socratic ignorance see the author's paper, "The Virtue of Socratic Ignorance," *American Philosophical Quarterly* 18 (1981).

6. The choice of terminology was partly influenced by Theodore Roszak's book *Person/Planet* (New York: Doubleday, 1978).

7. *Cartesianism* is used to refer to the historical influence of Descartes' philosophy. It is not entirely certain that Descartes would have approved the direction in which others developed his thought. For an illuminating philosophical and historical discussion of technique see William Barrett, *The Illusion of Technique* (New York: Doubleday, 1978).

8. James Wall, *The Titanic Effect* (Stanford: Sinauer and Associates, 1974).

9. *Humanism* as we use the term refers to that anthropocentric and secular philosophy which accompanied the development of science and technology. It emphasizes that science and technology can solve most human problems, that problems not solvable by technology can be solved by social engineering, that human values alone are important, and that the world is a resource for humanity. There were and are religious strains of humanism which recognize human limitations and the need for spiritual development, but we do not refer to these here. For a critique of the technocratic version of humanism see David Ehrenfeld, *The Arrogance of Humanism* (New York: Oxford University Press, 1978).

10. See Kant's two works, *Perpetual Peace* and *The Idea of a Universal History on a Cosmopolitan Plan.*

11. Lewis Thomas, *The Lives of the Cell* (New York: Bantam, 1974).

12. Alfred North Whitehead, *Modes of Thought* (New York: Macmillan, 1938), p. 11. This passage is cited by Roland C. Clement in his brief but helpful article "Watson's Reciprocity of Rights and Duties," *Environmental Ethics* 1 (1979): 353–55. Compare here also Whitehead's

comment in *Science and the Modern World* (New York: Macmillan, 1925), p. 136: "Remembering the poetic rendering of our concrete experience, we see at once that the element of value, of being valuable, of having value, of being an end in itself, of being something which is for its own sake, must not be omitted in any account of an event as the most concrete actual something."

13. William O. Douglas, "Sierra Club vs. Morton Minority Opinion," reprinted in C.D. Stone, *Should Trees Have Standing?* (Los Altos: William Kaufmann, 1974), p. 75.

14. Doug Boyd, *Rolling Thunder* (New York: Delta, 1974), pp. 51–52.

15. See Naess, "Shallow and Deep Ecology": "The ecological fieldworker acquires a deep-seated respect, or even veneration, for ways and forms of life. He reaches an understanding from within, a kind of understanding that others reserve for fellow men and for a narrow section of ways and forms of life. To the ecological field-worker, *the equal right to live and blossom* is an intuitively clear and obvious value axiom" (p. 95f). See also Holmes Rolston, III, "Can and Ought We to Follow Nature?" *Environmental Ethics* 1 (1979): 7–30.

16. I owe this point to Mark Sagoff, "On Relating Philosophy to Environmental Policy" (unpublished).

17. James Lovelock and Sidney Epton, "The Quest for Gaia," *New Scientist* (6 February 1975): 304–306.

18. This is put far too tersely. An insightful discussion of this can be found in David Bohm, *Fragmentation and Wholeness* (Jerusalem: Van Leer, 1976) and Ken Wilber, *No Boundary* (Los Angeles: Center Publications, 1979), esp. chapter four.

19. For a semipopular discussion of these metaphysical implications in relation to theoretical physics see Gary Zukav, *The Dancing Wu Li Masters: An Overview of the New Physics* (New York: William Morrow, 1979). To date however, A.N. Whitehead's *Process and Reality* remains one of the most sophisticated versions of the metaphysical implications of twentieth-century knowledge of the universe, as based on physics and biology. Whitehead's metaphysics attempts to take the process view of reality seriously, whereas many popular versions (Zukav's excluded) actually fall back on classical, atomistic materialism, which fails to account for the developmental character of the world, and which treats matter as inert stuff rather than as energy. Whitehead's philosophy recognizes the dialectical character of process, the interpenetration

of opposites, the significance of levels of organization, the importance of community, and the irreducible character of awareness. All these features give his organicism a particular relevance to the deep ecology movement.

This paper is dedicated to Justice William O. Douglas. An earlier version was presented as the first lecture in the "William O. Douglas Distinguished Lectureship" series at Yakima Valley College, Yakima, WA, May 1979. It is part of a larger project begun in the academic year 1974–75 and supported by the Canada Council and the University of Victoria. The author thanks Ms. Delma Tayer of Yakima Valley College for her efforts in organizing the lecture series, and Holmes Rolston, III, whose helpful criticisms and suggestions have greatly improved the paper. The analysis has also benefited from reading an unpublished paper by Donald A. Crosby, Colorado State University, "Authority in Social Systems: Two Models."

9

The Ecological Self

Bill Devall

When sociologists discuss *self* they usually are referring to the social self. When asked "who are you?," most people respond by saying: "I am a Christian" or "I am a male" or "I am a carpenter" or "a mother" or "an American." Sometimes people say, "I am an environmentalist." A person expressing ecological self would say, "I am a forest being."

We identify our social self with our social identity—occupation, gender, religion. And many people evaluate themselves based on perceived criteria concerning what their reference group says is important for their social identity.

While social scientists limit discussion to the social selves, many religious teachers have written about the "oceanic self," the mystical union with the One or with Wholeness or the Godhead. In the Christian tradition, the classic advice of an unknown monk in *The Cloud of Unknowing*, the writings of St. John of the Cross, and the more contemporary writings of Thomas Merton explore the process of unveiling the oceanic self through practicing prayer, contemplation, and meditation. While many Christian saints encourage believers to cultivate the oceanic self, St. Francis of Assisi in the thirteenth century seemed to encourage Christians to cultivate some aspects of their ecological self, at least to the extent of asserting a kind of deep ecology position (White 1967).

Exploring our ecological self is one aspect of "all around maturity," as Naess says. Naess' conception of maturity could also be called "full maturity" or "many-sided maturity." Humans are many-faceted beings and a person might be quite sophisticated and mature in professional activities but quite immature at social relationships. Or a person could be mature as a member of a family, having broad and deep empathy with spouse, children, and other kin, but not be mature politically. A person can also be mature in social relations but have an adolescent ecological self. Such a person might be very careless or unaware or insensitive to forests or rivers or, generally speaking, to "the land."

In his Keith Roby lecture at Murdoch University in 1986 [reprinted in this anthology], Naess said: "We may be said to be in, of, and for nature from our very beginning. Society and human relations are important, but our self is richer in its constitutive relations. These relations are not only relations we have with other humans and with the human community," they are relations we have with our home bioregion, or with plants and animals which co-inhabit our living space (Naess, "Self-realization," 1986).

The dominant view in modern society is to define what is not me as "the other." When the other is a bioregion, a forest or a redwood tree, then it is a "thing," an object which can be manipulated by and for humans for narrow purposes. But deep ecology understands the "I" in relation to the "other." The human ecologist Paul Shepard, in a famous essay on ecology and humankind, calls the ecological self the "relatedness of self."

> Ecological thinking ... requires a kind of vision across boundaries. The epidermis of the skin is ecologically like a pond surface or a forest soil, not a shell so much as a delicate interpenetration. It reveals the self ennobled and extended rather than threatened as part of the landscape and the ecosystem, because the beauty and complexity of nature are continuous with ourselves. (Shepard 1969, 2)

When we find our self only in our narrow ego or with certain attributes of our body which are socially desirable ("I am beautiful," "I am athletic," etc.) then we underestimate our self-potential, that is, our ability to move in a self-realizing way.

The ego has been defined in various ways by different psychologists but at the minimum it is a collection of memories, fantasies, information, and images about who we are. It is what we think we are, not what we experience as our self. A healthy ego is a sign of mental health in some theories of psychotherapy, but egotistic identity is a transitional stage from pre-egoic (early childhood) to broader existential identity of more mature persons. Clinging to rigid ego identity can be pathological and can include denial, projection, and repression (see Wilbur 1980, 1981).

The ego can be understood as the voice of the self, but when we use the ego to build a barbed wire fence around our feelings, to deny our vulnerability and deny our interconnections with watersheds, forests, and rivers, then the ego becomes a prison guard and not a voice.

From the perspective of transpersonal psychology, we are more richly alive when exploring patterns of relationships and interactions which we can call the self. Drawing upon the work of theorists such as Ken Wilbur, Frances Vaughan, a therapist and author, defines self as "an open living system in an intricate web of mutually conditioned relationships," as does Drengson (1986). Vaughan continues,

"This view recognizes both our biological and our psychological dependence on the environment. Although we may feel subjectively separate from nature and each other, we are actually interdependent and interconnected with the whole fabric of reality. We are conditioned by society and the environment at every stage in the evolution of self-concept. Yet we also shape the environment and co-create the social fabric that supports us. The complex web of relationships within which we exist involves a continuous flow of mutually determined interaction for which we can begin to take more responsibility as we understand our part in co-creating it" (Vaughan 1986, 33).

When a person stops defending an old ego identity—an image of oneself which does not correspond to current experiences—and disidentifies with his or her rigid social identity, growth can occur in the transpersonal self. The inner search for self-identity requires acceptance of our psyche and our physical vulnerability. It does not mean narcissistic self-absorption in our problems.

Exploring the ecological self is part of the transforming process required to heal ourselves in the world. Practicing means breathing the air with renewed awareness of the winds. When we drink water we trace it to its sources—a spring or mountain stream in our bioregion—and contemplate the cycles of energy as part of our body. The "living waters" and "living mountains" enter our body. We are part of the evolutionary journey and contain in our bodies connections with our Pleistocene ancestors. Extending awareness and receptivity with other animals and mountains and rivers encourages identification and engenders respect for and solidarity with the field of identification. This does not mean there will never be conflicts between the vital material needs of different people or between some humans and some other animals in specific situations, but it does mean that a basis for "good actions" or "right livelihood" is not based only on abstract moralism, self-denial, or sacrifice.

When exploring our ecological self openly and with acceptance, no judgment is made, nor is there a pursuit of anything. The self is not an entity or a thing, it is an opening to discovering what some call the Absolute or in Sanskrit, *atman*.

Awakening the self beyond the barbed wire fence the ego has constructed engages us in the world, in the grounding of being-in-the-world. Naess frequently talks of spontaneous joy we experience because we are part of what really is. Other aspects include compassionate understanding, wisdom, receptivity, intuitiveness, creativity, allowing to happen, connectedness, openness, and peacefulness. The expanded, deepened self is not impersonal but transpersonal.

The process of healing, in transpersonal psychology, begins with self-awareness. "We must know ourselves and accept ourselves before attempting to manipulate ourselves for the purpose of changes we think are desirable. To understand all is to forgive all, and forgiving ourselves for being just as we are is the first step to healing" (Vaughan 1986, 56).

As we discover our ecological self we will joyfully defend and interact with that with which we identify; and instead of imposing environmental ethics on people, we will naturally respect, love, honor, and protect that which is of our self.

> We need environmental ethics, but when people feel they unself-lessly give up, even sacrifice, their interest in order to show love for nature, this is probably in the long run a treacherous basis for conservation. Through identification they may come to see their own interest served by conservation, through genuine self-love, love of a widened and deepened self. (Naess, "Self-realization," 1986)

There are many reasons to defend the integrity of a landscape from the invasion of industrial civilization. Supporters of deep ecology especially defend the integrity of native plants and animals living in their own habitat unmolested by humans. We also defend the integrity of certain places (the Grand Canyon comes to mind) because they are awesome, beautiful and unique, or because we use those areas for sport or recreation.

Another strong reason is "if we, after honest reflection, find that we feel threatened in our innermost self. If so we more convincingly defend a vital interest, not only something out there. We are engaged in self-defense and defend fundamental *human* rights of vital self-defense" (Naess, "Self-realization," 1986). No moral exhortation or dogmatic statement of environmental ethics is necessary to show care for other beings—including rivers or mountains—if our self in this broad and deep sense embraces the other being.

Naess contrasts this view of enlightened self-interest with altruism which, he says, "implies that *ego* sacrifices its interests in favor of the other, the *alter*. The motivation is primarily that of duty: it is said that we ought to love others as strongly as we love ourselves" (Naess, "Self-realization," 1986). But humans have limited ability to love from mere duty or moral exhortation.

Spokespeople for the reform environmental movement, and some philosophers, offer many statements of environmental ethics and call upon people to sacrifice for future generations, for people in developing nations, etc. Now, according to Naess, "we need the immeasurable variety of sources of joy opened through increased sensitivity towards the richness and diversity of life." We need to be reminded of our moral duties occasionally, but we change our behavior more simply with richer ends through encouragement. Deeper perception of reality and deeper and broader perception of self is what I call ecological realism. That is, in philosophical terms, however important environmental ethics are, ontology is the center of ecosophic concerns.

John Seed provides an example of the joy and sense of empowerment that occurs when exploring our ecological self. Seed is a rainforest activist living in northern New South Wales, Australia. He participated in campaigns directed at stopping logging of remaining old-growth, subtropical rainforests in New South Wales in the late 1970s and early 1980s. With his growing awareness of his connection with rainforests, he organized the Rainforest Information Center, an educational organization dedicated to preservation of rainforests around the globe. He writes about his defense of rainforests in terms of his own psyche and his extended self in an essay entitled "Rainforest and Psyche."

> I believe that contact with rainforest energies enlivens a realization of our actual, our biological self. They awaken in us the realization that it was 'I' that came to life when a bolt of lightning fertilized the chemical soup of 4.5 billion years ago; that 'I' crawled out of Devonian seas and colonized the land; that, more recently, 'I' advanced and retreated before four ages of ice.
>
> ... our psyche is itself a product of the rainforests. We evolved for hundreds of millions of years within this moist green womb before emerging a scant five million years ago, blinking, into the light.
>
> If we enter the rainforest and allow our energies to merge with the energies that we find there, then the rainforest may be a place where our roots are able to penetrate through the soft soil reaching beyond the sad 16,000-year history and into the reality of our billions-of-years-of-carbon journey through the universe. Various truths which had been heretofore merely 'scientific' become authentic, personal, and yes, spiritual. We may now penetrate to a truly deep ecology. (Seed, "Rainforest," 1985)

In his essay on anthropocentrism, Seed makes a more general statement. "When humans investigate and see through their layers of anthropocentric self-cherishing, a most profound change in consciousness begins to take place." People stop identifying exclusively with their humanness and begin a process of transforming their relationships with other beings and commitments to them." Seed says, "I am protecting the rainforest' develops to 'I am part of the rainforest protecting myself. I am that part of the rainforest recently emerged into thinking.' "This is enlightened self-interest. Seed concludes, "... there is an identification with all life" (Seed, in Devall 1985, 243).

Self-realization has no artificial boundary. Some people say they identify with Mother Earth or with the biosphere. Naess, however, prefers to use the term *ecosphere* because this term implies a broader definition of living beings.

A community with appropriate rituals, social mentors, languages, art forms and methods of education can facilitate exploration of the ecological self. When we use the phrase "deep, long-range ecology movement" we already have become self-conscious, and have defined a need which is not fulfilled in dominant institutions. Ecological-spiritual cultures will emerge, of necessity, as more and more people together engage the healing process of exploring ecological self.

The search for personal growth is found in the New Age movement as well as the deep ecology movement. New Age ideology, to many of its critics, remains human-centered. According to some theorists in the New Age movement, the next phase in development of human consciousness is the development of planetary consciousness. Humans will become the "eyes and ears" of Gaia—the Earth organism.

Gaia consciousness may transcend ego and even transcend narrow sectarian loyalties to ethnic group or to nationality. But such consciousness can also remain anthropocentric. The environment, in the writings of some admired New Age thinkers such as Teilhard de Chardin, is still an environment of which humans are stewards. The narrow needs of some humans are still served before the vital needs of other species of plants and animals.

This kind of New Age ideology uses some symbols of nature and uses ecology as a slogan; but this ideology continues to insist that the best civilization is a technocratic (euphemistically called high-tech) civilization, and that space programs, human domination of evolutionary processes through genetic engineering, and complex computer systems, including massive computer modeling of ecosystems, is the most joyful future for humans on this Earth.

Besides the emphasis on technocratic manipulation of living processes, some New Age thinkers seem to emphasize a synthetic integration of different spiritual traditions, such as Native American spirituality and various group therapy processes, into a false sense of self-awareness. This New Age spiritual consciousness is comparable to the "false consciousness" which Marxists assert arises in late capitalist societies due to the incessant propaganda of the ruling class.

Ecological self is not a forced or static ideology but rather the search for an opening to nature (Tao) in authentic ways. If a person can sincerely say after careful self-evaluation and prayer that "this Earth is part of my body," then that person would naturally work for global disarmament and preservation of the atmosphere of the Earth. If a person can sincerely say, "If this place is destroyed then something in me is destroyed," then that person has an intense feeling of belonging to the place.

Sometimes resignation prevails when a person feels hurt by the destruction of primeval redwood groves or by destruction of rainforests in South America. People have some desire to help the rainforest as part of their extended self but say, "I can't do anything," or "That's progress." They feel guilty or depressed.

When our joyful sense-of-place includes the whole Earth we may also feel overwhelmed at the enormous task of defense, but this is a passing feeling. Gary Snyder in an article titled "Saving the Little Waterhole We Sing By" explains it thus: "We must know that we've been jumped, and fight like a raccoon in a pack of hounds, for our own and all other lives."

Cultivating ecological self does not mean learning how to use the joy stick in "spaceship earth" but how to joyfully blend with the watershed in which we live. In other words, it encourages modesty instead of hubris.

Naess comments, however, "Modesty is of little value if it is not a natural consequence of much deeper feelings. . . . The smaller we come to feel ourselves compared to the mountain, the nearer we come to participating in its greatness. I do not know why this is so" (Naess, "Modesty," 1979).

Examples of persons who explored their ecological self include John Muir who tramped the Sierra Nevada in the nineteenth century, Aldo Leopold in the wilderness of the American Southwest in the 1930s, poet Robinson Jeffers on the Big Sur coastline of California, and Gary Snyder dwelling in the foothills of the Sierra Nevada. Jeffers carries the process of self-identification into the realm of the oceanic self when he writes of "falling in love outward." He searches for the "tower beyond tragedy"—the tragedy of human civilization destroying rainforests and Jews and homosexuals and prairies and even whole cities in the twentieth century.

Total identification with "organic wholeness" is possible only after identification with some living being more immediate and tangible. Jeffers himself describes this process in one of his poems evoking a walk in the mountains of Big Sur:

I entered the life of the brown forest,
And the great life of the ancient peaks, the patience of stone,
I felt the changes in the veins
In the throat of the mountain, and, I was the stream
Draining the mountain wood; and I the stag drinking; and I was
    the stars,
Boiling with light, wandering alone, each one the lord of his own
    summit;
and I was the darkness
Outside the stars, I included them. They were a part of me.
I was mankind also, a moving lichen
On the cheek of the round stone . . . they have not made words for
    it, to go behind things, beyond hours and ages,
And be all things in all time, in their returns and passages in the
    motionless and timeless center,

In the white of the fire ... how can I express the excellence I have
found, that has no color but clearness;
No honey but ecstasy; neither wrought nor remembers; no
undertone nor silver second murmur
That rings in love's voice ...

Robinson Jeffers, *Not Man Apart,*
Lines from Robinson Jeffers (1965)

We identify with intermediate landscapes more readily than with
remote or abstract ones. The more we know a specific place intimately—know its moods, seasons, changes, aspects, native creatures—the more we know our ecological selves.

It is characteristic of most primal (not primitive as some call these
cultures in a pejorative pseudo-evolutionary mode of talking) peoples to consider spirit and matter as interwoven. The Koyukon of
Alaska, for example, live in a world that watches. "The surroundings are aware, sensate, personified. They feel. They can be offended.
And they must, at every movement, be treated with respect" (Nelson 1983). The central assumption of the Koyukon worldview is
that the natural and supernatural world are inseparable; each is
intrinsically a part of the other. Human and nonhuman entities are
in constant spiritual interchange.

Contemporary ecophilosophers recall Native Americans as exam-
ples of maturity, of having broader and deeper relations with *place.*
Listen, for example, to this statement by Chief Standing Bear of the
Oglala Sioux:

> Kinship with all creatures of the earth, sky and water was a real
> and active principle. For the animal and bird world there existed
> a brotherly feeling that kept the Lakota safe among them and so
> close did some of the Lakota come to their feathered and furried
> friends that in true brotherhood they spoke a common language.
>
> The old Lakota was wise. He knew that man's heart away from
> nature becomes hard; he knew that lack of respect for growing,
> living things soon led to lack of respect for humans too. So he kept
> his youth close to its softening influence. (McLuhan 1971, 6)

When the Native American says: "What a man does to the earth, he does to himself," we understand that the self he is speaking about is not the minimal self but the *Great Self* (see Hughes 1983).

The Minimal Self

The great nihilistic process of modern civilization, as Max Weber succinctly put it nearly a century ago, has been "the disenchantment of the world," and the rise of bureaucratic domination, mostly in the great urban centers of modernism. When a person says, "I am a New Yorker" or "I am a Californian" or when a person says, "I am a bureaucrat with Z agency" (CIA, FBI, IRS, FAA, USPS, USFS) they have allowed their broad self to be diminished by a bureaucratic identity.

Under the influence of philosophical assumptions of modern science, experts on nature—biologists, zoologists, soil scientists, wildlife managers, foresters, mammalogists—treat nature only as abstracted, objectified data. They kill their positive feelings of identification in order to be detached and neutral (see Merchant 1980; Berman 1981).

At the College of Natural Resources at my own university, professors argue endlessly about various models for codifying ecosystems, forest types, soil types, etc. They develop elaborate models using the fastest computers available to describe forests and oceans. But the forests remain "out there." Students are never encouraged to find a part of some forest and learn from it through emotional as well as intellectual experience. Students are taught to be objectively neutral to the forest. To be otherwise is to be labeled a sentimentalist or, worse still, an environmentalist. Students in natural resources sciences and management, therefore, are much like the guards in Nazi death camps. Their neutrality toward forests or wildlife or fish kills any natural feelings of empathy or sympathy they might have. If emotional responses to place and spiritual awareness are killed, and if all nature is just "dead matter," then the bureaucracy can work its will on the land without having to meet the expectations of the will-of-the-land.

The Hopi have a term, *Koyannisgasti*, which can be translated roughly, "life out of balance." The Hopi, a tribe dwelling in the south-western region of North America, worked diligently over hundreds of years to keep right relations with earth, sky, gods, and their own community of mortal humans. Hopi cosmology is a form of ecosophy and represents great ecological sensitivity and artistic expression of the embeddedness of humans in deep nature.

We contrast Hopi cosmology with modern lifestyles and the emphasis on narrow ego. When city dwellers are asked to locate their self, many point to their head or larynx. We talk much, listen very little. If we identify our self with our body, we worry about our body image and work diligently with weights or other exercises to build up our bodies to make them more appealing to other people.

The social critic Christopher Lasch calls this self-centered contemporary self-obsession with ego-gratification, social status, and the pursuit of hedonistic pleasures the "minimal self" (Lasch 1984, 16). The minimal self has contracted to a defensive core concerned primarily with psychic survival and making a good impression on certain significant people—bosses, clients, potential sexual partners. Seeing the problems of living in modern times—crime, increasing air and water pollution, terrorism, long-term economic decline, nuclear arms race, cynicism in major institutions in society—the minimal self prepares for the siege, retreats to private pleasure domes and withdraws from community service or any form of commitment to the peace movement or environmental movement.

Rootless, alienated from human community and from wild nature, from the will-of-the-land, besieged with propaganda from scientists and natural resources corporations insisting that humans can manage forests or rivers to improve productivity and that nature should be controlled by human technology, the goal of the minimal self is survival, not personal growth.

This constriction into the minimal self in many people has been linked to complex and interrelated changes in European societies during the past 2,000 years, to the rise of bureaucratic domination and to dominant Judaeo-Christian traditions, especially since the Protestant reformation. The geographer Tuan, in his book *Segmented Worlds and Self*, asserts that St. Augustine in the fourth century A.D.

was one of the first truly modern men because he was self-conscious, he wrote an intimate autobiography of his self-questioning, and he publicly confessed his most intimate tensions (Tuan 1982, 157).

But only with the close of the Dark Ages and the coming of the Renaissance in the fifteenth century, says Tuan, did the individual, isolated in a relatively small area or region, strive for segregation. Tuan traces the development of closed and segregated living spaces— separate rooms for toilet, sleeping, eating, holding formal parties, recreation—and suggests a correlation between segmented living spaces and segmented aspects of self-identity.

Intellectuals began to see themselves as detached observers rather than active and involved participants in all aspects of community life, and the distinction between public life and private affairs—love affairs, fantasies, dreams—became more distinct.

European intellectuals, the Catholic church, as well as economic systems, helped desacralize nature and thus opened the fields, forests, and mountains to almost unlimited economic use by humans, use restrained only by technological ingenuity and limited capital for investment. Land stewardship was turned into land management for maximum crop production (see Polanyi 1944).

For Tuan, the hallmark of modern mentality is subjectivism, the discovery of intellectual perspectives or opinions. Instead of finding a basis for group consensus by a process of discussion, intellectuals, especially, found a particular opinion to defend and upon which to discourse. Reputations were made or lost on the basis of wit and intellectual banter rather than on authentic spiritual questing.

The social self became a series of presentations. Social psychologist Irving Goffman uses the imagery of the theater to express how individuals become players on a team. Presentations are staged for the benefit of an audience which evaluates the performance. Work places, restaurants, offices, and so forth are divided into front regions where the players perform, and back regions where they prepare for their presentations (Goffman 1959).

The minimal self, with its desperate defenses against any feelings of respect or love or any values other than materialism and success as measured by expensive automobiles, houses, computers, or whatever is a status symbol, is a very unhappy self. Many people who are successful in careers in their 30s and 40s become extremely unhappy. Satisfaction from material rewards diminishes with age, and without any of the traditional bonds of community and lacking a sense of extended self-identity, the minimal self is entrapped and rebellious.

If children are taught to be only "New Yorkers" or "San Franciscans" they identify primarily with a human-built environment. Biological diversity is greatly diminished in cities. When children see wild animals only in a zoo, they see imprisoned animals, not animals in their natural habitat with predator and prey dancing their dance. The "organic community" described by sociologists and historians is replaced by the "mechanical community." The ultimate of this process is the cruise through the computerized swamp at amusement parks with plastic alligators and other creatures which attack and retreat on cue.

The real, organic community is simple in material goods but rich in individuation, communalism, awareness of the way things are, in affectional and spiritual connections with a specific landscape.

Is it possible to explore our ecological self while imprisoned in the concrete streets of a modern metropolis?

Martin Krieger suggests that the urban landscape of the future might be one of plastic trees and flowers. "Why not plastic trees?" Krieger asks. They require less maintenance than real trees. They are always green but don't need water. They can seem like real trees (Krieger 1973). Although not including plastic trees, some futurists suggest that we can create an ecosystem in a space station complete with plants and animals if we wish. And some visionary engineers suggest that by putting huge plastic bubbles over our cities, we can create rainforests. Such proposals are logical extensions of the dominant way of thinking in our present culture and they demonstrate that a radical shift in self-awareness is imperative.

Let there be no mistake about our situation. The enormity of human population requires that most people live in big cities, but sound ecological policies could make cities *somewhat* more livable for the all-around well-being of humans and all other living beings.

Even in the concrete depths of the largest city, a person can explore the bedrock upon which the city is built and trace the watersheds of streams and rivers channeled in concrete pipes. A person can feel the suffering of city-dominated watersheds and work for reconciliation.

We can also experience one aspect of nature which is a lesson in humility, if not modesty, in large cities. We can experience great storms which newscasters like to call natural disasters—wind storms, hurricanes, heavy snow storms, typhoons, cyclones. Humans sometimes die in these storms; travel is disrupted and property is destroyed, but frequently only because humans did not read the land appropriately and they built their houses or other structures too near an ocean or river. Building towns on barrier islands of the southeast coast of North America, for example, is an invitation to disaster for humans because these towns lie in the path of hurricanes.

When we have empathy and solidarity with beings who don't reciprocate in our own terms, we gain in richness of experience. We know when a baby reciprocates our affection or negative feelings even though we don't talk with the baby except by cooing, singing, etc. We even know when certain mammals have a relationship with us. Paul Shepard in his book *Thinking Animals* argues that we have a need to form such relationships. Our relationships with domesticated dogs and cats are extremely important to many people, but they are pale and one-dimensional in comparison to the rich relationships our ancestors had with many kinds of animals in the wild. If we treat our domesticated animals as subservient then we only identify with them as masters, using them to give us satisfaction (see Shepard 1978).

Exploring ecological self can be partly described as discovering a sense-of-place or an ecological consciousness. Thus the more we know a place intimately, the more we can increase our identification with it. The more we know a mountain or a watershed, for example, and feel it as our *self*, the more we can feel its suffering. Some people tell me they feel the whole planet suffering during the present era. They tell me that if some of the vital organs of the planet are suffering, they want to sit with the planet just as volunteers for Hospice sit with a dying person, empathizing with that person. This is not an attempt to anthropomorphize natural entities or the planet, but an acceptance that "living beings" has a broader meaning than we usually ascribe to that term. We cannot know intimately all the suffering of a dying person and we do not know intimately all the connotations of "living being" when we use that term to describe the planet or a rainforest, for example. But without sentimentality or romanticism we can appreciate the possibility that living beings suffer and thus have empathy with them.

The positive message of [Naess' support for] deep ecology is maximal Self-realization of all beings, not just human beings and not just a narrow sense of personal growth. "Self-realization," says Naess, "in its absolute maximum is ... the mature experience of oneness in diversity.... The minimum is the self-realization by more or less consistent egotism—by the narrowest experience of what constitutes one's self and a maximum of alienation. As empirical beings we dwell somewhere in between, but increased maturity involves increase of the wideness of the self" (Naess, "Identification," 1985, 261).

Beyond Borders: Experiencing Bioregion

"In a life span, a man now—as in the past—can establish profound roots only in a small corner of the world." (Tuan, *Topophilia*)

Ecological self seems most accessible to us not by focusing on human-built places or on the organic whole or Gaia initially, but on our own bioregion.

Bioregion is a term combining *life* (bio) and *territory* (region or area understanding). The origin of the term is unknown, but the Canadian poet Allen Van Newkirk used the term in 1974 seeking to link the study of cultural and biotic regions. He also referred to bioregion as a point of view. Van Newkirk spoke of "bioregional strategies" for restoration of the Earth's natural plant and animal diversity within a "regional framework" and of cultural adaptation to specific bioregions. Ecology, language studies, poetry, myth, and cultural history are tools to be used in bioregional studies (Van Newkirk, 1975; Parsons 1985).

The central question for bioregional studies is what information do I need to know in order to live rightly and appropriately in this place?

Jim Dodge, a northern California rancher and writer, points to the intersection between bioregion and self-identification in a perceptive essay on bioregional theory and practice:

> To understand natural systems is to begin an understanding of the self, its common and particular essences—literal self-interest in its barest terms. "As above, so below," according to the old traditional alchemists; natural systems as models of consciousness. When we destroy a river, we increase our thirst, ruin the beauty of free-flowing water, forsake the meat and spirit of the salmon, and lose a little bit of our own souls. (Dodge 1981, 7)

Bioregion is discovered by each person through physical encounters with rivers or with whole watersheds of the region, the "veins of the landscape" as some call them. Cultural history may give some clues of appropriate relationship for contemporary citizens of a bioregion.

Four criteria for exploring a bioregion have been suggested by Dodge. The first is based on *biotic shift*, or the percentage change in plant and animal species from one region to another. If there is a 15–25 percent change in plant and animal communities then a biotic shift has occurred. With a shift in biotic communities there is likely to be a shift in land form, climate, and soils. There is much debate among scientists over the percentage of change which constitutes a biotic shift. Personal discoveries, however, are more important in exploring our "terrain of consciousness" than definitions given by scientists.

A second possible basis for bioregion is *watershed*. Indeed, some watersheds define basic hydrological units. By following our watershed we can trace the water we drink to its source. Of course, certain very large rivers, such as the Mississippi or Amazon, may lose their bioregional character, but intradrainage distinctions can be made. . . .

A third possible criterion for bioregional consciousness is our awareness of the *spirit-of-place*, or sense-of-place. Rediscovering sacred places in a bioregion might arise out of the common agreement within a culture, perhaps inspired by a religious leader. Although the origins of the designation of sacred place remain secret or hidden, the place is special within the human community.

> Sacred places are the sites for ceremony and ritual healing, contemplation, and rites of passage. They are honored in song and dance, through myth, symbol, and metaphor and become a catalyzing force for the celebration of nature and life itself. . . . In many cultures, such sacred places were seen and are seen today as the very cornerstone for cultural renewal. (Swan 1983, 33). . . .

A fourth criterion for bioregion is *cultural distinctiveness*. Rituals, art forms, distinctive ways of living and specialized terminology referring to land forms or weather or relationships with the landscape may indicate a biogeographical culture. . . .

Bioregional movements are political and social expressions of our vital need to be part of, not apart from, the place wherein we dwell. The North American Bioregional Congresses brought together people from many bioregions to discuss philosophy and practical skills for making a home in their own bioregions. The following statement from the preamble to the first North American Bioregional Congress in 1984 presents statements which are consistent with the deep ecology movement. . . .

> A growing number of people are recognizing that in order to secure the clean air, water and food that we need to healthfully survive, we have to become stewards of the places where we live. People are discovering that the best way to take care of ourselves, and to get to know our neighbors, is to protect and restore where we live.

Bioregionalism recognizes, nurtures, sustains, and celebrates our local connections with land; plants and animals; river, lakes and oceans; air; families, friends and neighbors; community; native traditions; and systems of production and trade.

It is taking the time to learn the possibilities of place.

It is a mindfulness of local environment, history, and community aspirations that can lead to a future of safe and sustainable life. It is reliance on well-understood and widely used sources of food, power, and waste disposal.

It is secure employment based on supplying a rich diversity of services within the community and prudent surpluses to other regions.

Bioregionalism is working to satisfy basic needs through local control in schools, health centers, and governments.

The bioregional movement seeks to recreate a widely shared sense of regional identity founded upon a renewed critical awareness of and respect for the integrity of our natural ecological communities.

People can join with neighbors to discuss ways we can work together to 1) Learn what our special local resources are, 2) Plan how to best protect and use those natural and cultural resources, 3) Exchange our time and energy to best meet our daily and long-term needs, and 4) Enrich our children's local and global knowledge.

Bioregionalism begins by acting responsibly at home. Welcome home! . . .

Summary

What will be the upshot if we embark on a voyage of discovery of ecological self? One writer asks if an active political stance is any longer appropriate with a broad ecological self. Will we come to a zenlike acceptance that we, individually, are part of nature—both in all its creative and destructive aspects? Nature survives. Why be a conservationist? Nature destroys as well as creates. It destroys more efficiently than humans with floods, fires, volcanic eruptions, meteor impacts, ice ages, and earthquakes. Humans in their nature certainly have aggressive tendencies as well as loving, caring tendencies. If we accept ourselves in all the fullness of living and if we accept that ultimately nature is indestructible by human agents, then it no longer matters if we conserve any particular aspect of nature.

Persons with a deep ecological understanding affirm the integrity of nature in the widest sense, yet we talk incessantly about the "death of nature"—of species extinction, rainforest and wetland destruction—due to human interventions. We stress the destructive and aggressive, shortsighted tendencies of some humans. When we emphasize these tendencies in our writing, speeches, and calls for action, we might frighten many ordinary people who will engage in psychological denial as a defense of their minimal self. They literally close their eyes to toxic wastes, deforestation, and other environmental ills.

And what of those people who cling so desperately to their narrow, defensive ego they have constructed in a pathetic attempt to defend themselves from annihilation in mass society? We can carefully demonstrate that identification with the wider circle does not mean we lose our individuation. All identification is relative. We maintain a relative ontological individuation while understanding our functional unity and relationship with the place in which we dwell. When my identity is interconnected with the identity of other beings then my experience and my existence depend on theirs. Their interests are my interests.

Beings are alive, in the widest definition of aliveness, and are ruled by the principle of self-realization, that is, an impulse for self-maintenance, self-preservation, and self-increase or self-perfection, to persist in their own being. Nature in the broadest sense is a self-realizing, internally interconnected cosmos. In the metaphysics of deep ecology [as an ecosophy], then, it has a will-to-live.

If I, as a part of this wider connection, also have individuation and a will-to-live, then I can act, should act, in self-defense when my broader and deeper interests are threatened. Conservation is thus self-defense. When I identify with primeval forests of redwood trees I want to defend them from logging because they are part of my sense-of-place.

However, my life is only a tiny breath in the cosmos. What does it matter if I do not conserve rainforests or redwood forests? But this begs the question of how should I live. If I express my own will-to-live then I contribute to the flourishing of aliveness in the cosmos, be it ever so small a flourishing in my own life among many lives. By expanding our will-to-live we contribute to realizing the biosphere more fully.

Since many people live only with a narrow awareness of self due to their cultural conditioning, it is most important in the deep, long-range movement to encourage the deeper ecological self to contribute to the flourishing of self-realization in the whole biosphere.

Humans have a unique, vital place in the natural order, but that place is realized not through technological manipulation but through participation in rituals and emotions, thoughts, prayers, and what Martin Heidegger calls the "round dance of appropriation."

This is not to imply that humans are most important, or even have a central place in cosmic self-realization. Humans can enhance the cosmic will-to-live, self-realization, in a small but vital way. Nature needs us as life-affirming people. Affirmation of our own self-realizing ecological self embraces more and more of the "other" into ourself. The more open, receptive, vulnerable, adventurous we are, the more we affirm the integrity of being-in-the-world.

In order to extend our *self* from the minimal self to more mature, maximum self-realization, we need more self-conscious awareness in our daily lives.

Many of us never take the time to engage in this process of discovery. We let ourselves become colonized by mass media, by expectations in our culture. We are seduced by entertainments and promises of pleasures on city streets. We break away only by becoming self-conscious. Thus we have a paradox, in order to lose our *self* into the larger self, we must become more self-conscious in the midst of techno-scientific civilization. Without cynicism or sentimentalism, we create an opening for discovery. Outside the ordered, bordered, fenced, domesticated, patrolled, controlled areas of our region, our wild self is waiting.

References

Berman, Morris. *The Reenchantment of the World*. Ithaca: Cornell University Press, 1981.

Dodge, Jim. "Living by Life: Some Bioregional Theory and Practice." *CoEvolution Quarterly* (Winter 1981): 6–12.

Drengson, Alan. "Developing Concepts of Environmental Relationships." *Philosophical Inquiry.* No. 8 (1986): 50–65.

Goffman, Erving. *The Presentation of Self in Everyday Life.* Garden City, NY: Doubleday, 1959.

Hughes, J. Donald. *American Indian Ecology.* El Paso, TX: Texas Western Press, 1983.

Jeffers, Robinson. *Not Man Apart, Lines from Robinson Jeffers.* Edited by David Brower. San Francisco: Sierra Club Books, 1965.

Krieger, Martin H. "What's Wrong with Plastic Trees?" *Science* 179 (February 1973): 451.

Lasch, Christopher. *The Minimal Self: Psychic Survival in Troubled Times.* New York: W.W. Norton, 1984.

McLuhan, T.C., comp. *Touch the Earth: A Self-Portrait of Indian Existence.* New York: E.P. Dutton, 1971.

Merchant, Carolyn. *The Death of Nature: Women, Ecology and the Scientific Revolution.* San Francisco: Harper and Row, 1980.

Naess, Arne. "Identification as a Source of Deep Ecology Attitudes." In *Deep Ecology.* Edited by Michael Tobias. San Diego: Avant Books, 1985.

Naess, Arne. "Modesty and the Conquest of Mountains." In *The Mountain Spirit.* Edited by Michael Tobias and Harold Drasdo. Woodstock, NY: Overlook Press, 1979: 13–16.

Naess, Arne. "Self-realization: An Ecological Approach to Being in the World." Murdoch University, 12 March, 1986.

Nelson, Richard. *Make Prayers to the Raven: A Koyukon View of the Northern Forest.* Chicago: University of Chicago Press, 1983.

Parsons, James J. "On 'Bioregionalism' and 'Watershed Consciousness'." *The Professional Geographer* 37 (February 1985): 1–6.

Polanyi, Karl. *The Great Transformation.* Boston: Beacon Press, 1944.

Seed, John. "Rainforest and Psyche." Lismore, New South Wales: Rainforest Information Center, 1985.

Seed, John. "Anthropocentrism!" in Devall, *Deep Ecology.* Salt Lake City, UT: Peregrine Smith, 1985: 243–246.

Shepard, Paul. "Introduction—Ecology and Man—a Viewpoint." In *The Subversive Science: Essays Toward an Ecology of Man.* Edited by Paul Shepard and Daniel McKinley. Boston: Houghton Mifflin, 1969: 1–10.

Shepard, Paul. *Thinking Animals: Animals and the Development of Human Intelligence.* New York: Viking Press, 1978.

Snyder, Gary. "Wild, Sacred, Good Land." *Resurgence* no. 38 (May/June 1983): 10–15.

Swan, Jim. "Sacred Places in Nature." *Journal of Environmental Education* 14 (Summer 1983): 32–37.

Tuan, Yi-fu. *Segmented Worlds and Self.* Minneapolis: University of Minnesota Press, 1982.

Tuan, Yi-fu. *Topophilia: A Study of Environmental Perception, Attitudes and Values.* Englewood Cliffs, NY: Prentice-Hall, 1974.

Van Newkirk, Allen. "Bioregions: Towards Bioregional Strategy for Human Cultures." *Environmental Conservation* 2 (1975): 108–109.

Vaughan, Frances. *The Inward Arc.* Boston: Shambhala, 1986.

White, Lynn, Jr. "The Historical Roots of Our Ecologic Crisis." *Science* 155 (1967): 1203–1207.

Wilbur, Ken. *The Spectrum of Consciousness.* Wheaton, IL: Theosophical Publishing House, 1980.

Wilbur, Ken. *No Boundary.* Boulder, CO: Shambhala, 1981.

10

Conservation and Self-Realization: A Deep Ecology Perspective

Freya Mathews

If we adopt the more "spiritual" approach to nature that deep ecology prescribes, will we in fact find that the kind of interventionist political action involved in conservation is no longer appropriate? Will we come to see that nature, in its widest sense, stands in no need of our "protection"? Does the same chest-beating self-importance lie behind the urge to conserve, guard, look after things, as lies behind the urge to blow everything sky-high? Is it the same human hubris, the same conviction that we are somehow outside of nature, that leads us to think that we can either destroy or "save" it? Does not destructiveness fall under the "law" of nature, just as much as creativeness does? Does not humanity the exploder and wrecker mirror nature just as faithfully as humanity the worshipful creator and conserver? Does deep ecology in fact require of us this kind of Zen-like surrender to the reality of our own aggressive tendencies, an acceptance of ourselves as a natural phenomenon—albeit a natural disaster perhaps—on a par with ice ages and interplanetary collisions, events which our Earth, Gaia, has incidentally succeeded in turning to the long-term advantage of life?

To see nature in this way, as able to take care of itself, as ultimately out of our destructive reach, poses a dilemma for conservationists: if we identify deeply enough with such an indestructible nature, seeing our Earth as a single manifestation of an infinite, inexhaustible principle, a cosmic principle of life, then this alleviates our angst and despair at the prospect of ecocatastrophe, because it means that ecocatastrophe does not spell the "death of nature" in its widest sense. On the other hand, such a deeper understanding of nature, which enables us to be life-affirming in the face of ecocatastrophe, seems to obviate the need for conservation, for if Gaia is but a single manifestation of a deeper and inexhaustible principle of life, then it no longer matters whether or not Gaia is preserved.

I raise this dilemma because I think that it is vital that we do not feed that soul-destroying angst and despair to which we as conservationists tend to be prey—and I am speaking of the despair that we feel not merely at the prospect of our own extinction, but at the awesome prospect of the end of nature, the end of the very Wheel of Birth and Death. I think that it is vital for our own spiritual health, and for the health of the environmental movement, that we reassure ourselves that we cannot bring about such a death of nature. Life-affirming people cannot be drawn to a movement which lives forever in the shadow of the belief that "the end is nigh"—and life-affirming people are, I try to show, the kind of friends that Gaia needs right now.

Indeed, I think that we cannot hope to change Western civilization, in the ways that we as conservationists dream of, if we constantly threaten it with the wholesale ecocidal consequences of its aggressive activities. From a psychotherapeutic point of view such threats are likely to bring about denial and repression rather than growth and enlightenment. Just as the young child needs to be reassured that its angry tantrums cannot really destroy the needed—and beloved—mother, so I think that Western civilization, at this stage, needs a degree of reassurance rather than terrorization. The child who is frightened of its anger represses and denies it, thereby giving it a life of its own and a license to act out its impulses in a literal and dissociated fashion. If Western society is too frightened of the consequences of its exploitation of nature, it will simply refuse to claim those consequences; it will banish its aggressive tendencies to the night, to unconsciousness, where they will be free to take their most literal, dissociated, and barbaric form. As psychotherapists always insist, you cannot break through to the next stage of consciousness unless you embrace your devil, and you cannot embrace your devil if you are too terrified of its destructiveness. Yet if you do not embrace it, but resist it, not only can you not break through, but you feed its power. The attitude that we may need to assume, as people seeking to shift the values of Western civilization, is thus not one of point-blank rejection of the evils embodied in this civilization, but a recognition of their origins in nature itself, and an acceptance and assimilation of them. This evil will then find its expression in integrated, perhaps symbolic, forms, and will not be left to act itself out in autonomous fashion.

The disadvantage of arguing that nature cannot be destroyed by human agency is, however, as I have said, that it appears to obviate the need for conservation. I think that deep ecology does indeed provide the reassurance that we need vis-à-vis the viability of nature, and protects us from despair, while at the same time providing an ongoing impetus for conservation—though we may find that it brings about a certain "deepening" of the meaning of conservation.

To begin at the beginning, then, let us approach that inevitable question, "What *is* deep ecology?" Saying what deep ecology is, is a bit like trying to say what Christianity is, or science, or Marxism: the basic principles that were initially set forth as the axioms of deep ecology[1] are so extraordinarily rich in meaning, and figure in so many traditions of thought, that they inevitably accrete different layers of meaning with each interpreter. I focus on that principle of deep ecology which is most relevant to the issue that I have just outlined—and which, in my view, constitutes the very heart of Naess' deep ecology philosophy—self-realization.

Deep ecology is concerned with the metaphysics of nature, and of the relation of self to nature. It sets up ecology as a model for the basic metaphysical structure of the world, seeing the identities of all things—whether at the level of elementary particles, organisms, or galaxies—as logically interconnected: all things are constituted by their relations with other things. How exactly this view is supposed to cash out within the framework of theoretical physics is still uncertain, though hints and clues abound. In the framework of biological ecology, however, the idea of interconnectedness is, of course, completely at home: our ecological definition, or identification, of a blue whale, for instance, will involve reference to krill, and our definition of krill may involve reference to the blue whale. From the viewpoint of the interconnectedness thesis, organisms are not logically constituted by their physical structure only, but by their relations to the elements of their environment: organisms are seen as essentially interactive beings, logically incapable of existing independently of other beings.

Deep ecology (Naess' Ecosophy T) takes this kind of interconnectedness as absolutely fundamental to the identity of things—as nonreducible to any kind of mechanistic base. This has the further consequence of rendering the relation between part and whole holistic as opposed to merely aggregative. Let me explain this. If the interconnected elements are seen as constituting a greater whole, then each element, being logically constituted by its relations with the other elements, is conditioned by the whole. The elements are given simultaneously with the whole, since they cannot be given independently. In this sense, the relation between part and whole is genuinely holistic, in that the parts constitute the whole, but are not given independently of it, i.e., the nature of the whole conditions the parts. Contrast this with the ordinary—atomistic—conception of the relation between part and whole. In the ordinary way of thinking, the parts exist independently of the whole which they aggregatively constitute: while the parts determine the nature of the whole, the whole in no way determines the nature of the parts.

It should be pointed out that interconnectedness does not imply that organisms do not possess a genuine individuality: their functional unity confers on them an essential ontological distinctness and integrity, but this individuality is strictly relative—it is itself a function of the particular environment which is capable of sustaining such a self-realizing, self-maintaining system. A relative ontological individuality, on the one hand, and interconnectedness, on the other, are thus not in this framework mutually exclusive; on the contrary, they entail each other.

Applying this principle of interconnectedness to the human case, it becomes apparent that the individual denoted by "I" is not constituted merely by a body or a personal ego or consciousness. I am, of course, partially constituted by these immediate physical and mental structures, but I am also constituted by my ecological relations with the elements of my environment—relations in the image of which the structures of my body and consciousness are built. I am a holistic element of my native ecosystem, and of any wider wholes under which that ecosystem is subsumed. It is accordingly part of my essence that I stand in certain relations to the relevant elements of my environment. Since this is part of my essence, I cannot be said to flourish, to actualize the potentialities of my nature—in a word, to be fully self-realized—unless I do stand in these relations. Moreover, in order for me to stand in these relations, and thus achieve self-realization, the elements in question must exist. Thus, it is in my interests to ensure that those elements exist.

Through such reasoning we may be led to a tenet of deep ecology [ecosophy T, e.g.] I call the *identification thesis*, according to which I identify with the wider systems of nature: if my identity is logically interconnected with the identity of other beings, then, as I have explained, my chances of self-realization depend on the existence of those beings.[2] It is therefore in my interest (viz., my interest in self-realization) to ensure their existence, and in this sense our interests converge. I identify with the wider systems of nature in the sense that I recognize their interests as my own.

There is a very important assumption involved in this line of reasoning, which is not universally accepted or acknowledged in the literature of deep ecology, namely, that the wider systems of nature do possess interests. That is to say, interconnectedness alone is a necessary but not a sufficient condition for the identification of self with nature. It is logically possible for a world of rocks to exhibit interconnectedness—perhaps matter, in this world, is just a particular type of topological disturbance in a substantival spatial continuum. In this case, a topological disturbance, or rock, in one region of space is "interconnected" with topological disturbances, or rocks, in other regions, in the sense that a configuration in one region of this elastic space helps to determine the configurations in the surrounding regions. But this "interconnectedness" does not entail the identity of a given rock with the rock world as a whole; nor does it make sense to say that one rock can or should "identify" with others, since the rock lacks interests, as does the rock world as a whole, and to "identify with" another is essentially a matter of it assuming the interests of the other.

This point raises the question of what it means to say that a being has interests. A being may be said to have interests if it has needs, and it may be said to have needs if it is seeking to maintain or realize its own existence—it *needs* those things which contribute to its self-maintenance or self-realization. A rock cannot be said to have interests because it has no investment in its own existence—it in no way resists causal inroads into its physical integrity. The rock possesses a contingent unity only—it waits passively to be ground to dust, or worn away by wind and rain. An organism, in contrast, is essentially a system for realizing and perpetuating itself, and the more complex it is the more efficiently it resists the forces of disintegration. To have interests, then, is to be ruled by the principle of *self-realization*, a principle to which medieval philosophers referred as the "conatus." The *conatus* was the impulse, not only for self-preservation or self-maintenance, but also for self-increase or self-perfection—an impulse that is present in all living beings. Spinoza defined it as the "effort by which each thing endeavors to persist in its own being."[3] Clearly the *conatus* includes the impulse to realize in full the potentialities of one's nature or logical essence as well as the impulse merely to survive on a physical level, i.e., to maintain a certain physical structure. It is in this fuller sense that it is in our interests to promote the interests of the wider systems of nature—and clearly the identification of self with nature is only possible when nature is itself possessed of *conatus*.

What justification can be offered for this apparently counterscientific idea of nature as animated by a principle of telos, or a will-to-exist? In many of the expositions of deep ecology [supporters] this idea is present merely as an implicit axiom, accepted on intuitive rather than rationalistic grounds. I think that some justification is called for, and I have attempted to construct one elsewhere.[4] As I do not wish to enter into that argument here, however, I shall just accept, for the purposes of our present discussion, that nature in its widest cosmic sense is indeed a being endowed with conatus, a self-realizing entity. I use the term ecocosm to characterize such a self-realizing, internally interconnected cosmos.

What then is the normative significance of the identification thesis? Does our identification with the widest possible whole—with the ecocosm—entail that we should practice conservation? This is certainly the implication that has been assumed to hold within the literature of deep ecology. It has generally been argued that my identification with the wider wholes of nature entails that I will defend them when they are under attack. Conservation is in this way seen to be purely a matter of self-defense.

Before we can evaluate this argument, we need to distinguish between different levels of nature, which are also, in the light of the identification thesis, levels of my self. Or rather, identification may be pictured in terms of a series of concentric circles: the center point is my immediate bodily self, the next circle my ecological self, and the final circle is my cosmic self. Looking at our question in the light of these widening circles of self, we ask: does achieving identification with the cosmic self cancel out my need to defend the more immediate circles of self—the bodily and the ecological? From the viewpoint of the ecocosm, does it matter whether local individuals and systems are conserved or not? If it does not matter to the ecocosm whether local individuals or systems are conserved, does my identification with the ecocosm imply that the fate of individuals and systems—including myself and my own system—does not matter to me? In other words, if I become sufficiently spiritual in my approach to nature, and see nature as a whole as my true Self, does this cancel out my natural impulse to defend and maintain my individual self—and hence too the ecosystem on which this self depends?

In order to answer this question, we have first to settle the question whether the fate of particular individuals affects the self-realization of the ecocosm. On the face of it, the answer appears to be no. Even when the cosmos is viewed—as, I have argued, deep ecology must view it—as a self-realizing being, which has to work at keeping itself in existence, it cannot be denied that the process of cosmic self-realization involves a continuous arising and passing away at the level of particular forms. Indeed the death and decay of particulars is obviously as intrinsic to this process as their reemergence in new forms: new life arises from the often violent destruction and deterioration of the old. Planets are born out of the shattered remains of blown-out stars. Elementary particles are constantly created and annihilated in the endless dance of energy at the subatomic level. If this is the "law" of the ecocosm, then when we adopt the ecocosmic viewpoint, the arising and passing away of individual lifeforms will not bother us, since it is consistent with the continued existence of our true—cosmic—self.

To argue in this way, however, is, I think, to misconstrue the identification thesis. Certainly there is a psychological sense in which I can "identify with" anything, or with any being with interests, so that the interests of that being displace my own. For instance, I could, in a psychological sense, identify with someone who wanted to kill me, and then it could be said that it would be "in my interest" for me to be killed, in the sense that I have taken on the other person's interests as my own. But such psychological identifications may be arbitrary or irrational—as the tendency to make bizarre identifications in psychosis shows. The identification thesis put forward in deep ecology (Naess' Ecosophy T) is rationally justified, and it supervenes on interconnectedness, as I explained earlier: I am only required to identify with a system, S, if my identity, and hence my self-realization, are bound up with S. In other words, the motivation for identifying with S is that S's interests complement mine. If it turns out that S's interests in fact conflict with mine, then this justification for the identification dissolves, for interconnectedness is not, as we saw, equivalent to complete identity: interconnectedness is just interconnectedness, no more.[5] Identification, in the context of deep ecology, is premised on a convergence of interests.

Hence, my essence, and therefore my chances of self-realization, are bound up with the fact that the ecocosm permits me to identify with it. The fact that my individual fate and the fate of my local ecosystem are not necessarily of great moment to the ecocosm does not mean that I should, on the strength of my cosmic identification, give up responsibility for self and ecosystem, for it was the interests of this immediate self which prompted the identification in the first place. My identification with the more immediate circles of self, viz., ecosystem and biosphere, thus requires that I defend and maintain them, and this self-defense argument is *not* nullified by identification with a cosmic whole that is tolerant of destruction at the level of particulars.

Although I have said that the ecocosm appears to be unscathed by the passing away of particulars, this is not to say that it is invulnerable, indestructible. It is, as a self-realizing being, like us, sustained by will, not cause, and will, unlike cause, is capable of faltering, failing. If the will falters, then the fabric of nature may indeed begin to unravel. However, the only way in which the cosmic will could be undermined by us would be if we experienced a radical failure in our own *conatus*, our own will-to-exist, for it is perhaps by way of our *conatus*, rather than through our physical fate, that we contribute to the cosmic flourishing, that we truly participate in the cosmic essence. If we developed a will-to-oblivion in place of *conatus*, as in the Arthurian myth of the wasteland,[6] then indeed might the universe begin to "sicken." (Note that will-to-oblivion is not the same as will-to-destruction. The latter is a form of will-to-power, which is in turn a form of will-to-exist.) It is clear, however, that whatever it is that is at present wrong with our culture, it is not our *conatus* which is at fault: our impulse toward self-maintenance is still robust. From the point of view of deep ecology, what is wrong with our culture is that it offers us an inaccurate conception of the self. It depicts the personal self as existing in competition with and in opposition to nature. Our *conatus* is accordingly unduly limited, and therefore cannot succeed either in its long-term aim of maintaining us in existence or in its immediate aim of self-realization. If we destroy our environment, we are destroying what is in fact our larger self. We commit this mistake because we are suffering from a maladaptation in the form of a faulty belief system that misrepresents our identity to us. Like many another species healthily eager to live and flourish, yet handicapped by a functional flaw, we are innocently selecting ourselves out of existence. In doing so we are not, as I have said, letting nature down, any more than deer with outsize antlers are. Nor are we even letting down our ecosystem, since in light of our interconnectedness, our dysfunction is really its, and it is through us that it is selecting *itself* out of existence. Since our maladaptation is correctable, however, in a way that outsize antlers are not, our *conatus* calls on us to rectify it, to change our belief system to serve our ultimate interests.

Although I have said that our *conatus* is clearly still functioning, it is important to recognize that *conatus* is a matter of degree. When our sense of self is diminished, then our sense that the self is worth preserving must also diminish. A strong *conatus* requires a strong, rich sense of self, which means a self encompassing the widest possible circles of being. The stronger our *conatus*, the greater our contribution to the flourishing of the ecocosm. Like any self-realizing being, the universe may achieve varying degrees of self-realization. At present the sense of self entailed by the world view of Western culture is clearly relatively cramped and poor. By correcting this world view, and expanding our *conatus*, we might be able to help to realize the ecocosm more fully.

Many traditional or primal peoples have shared this sense that our *attitude* to nature, to life, is important to the welfare of the cosmos, not merely in the sense that a sympathetic attitude brings about ecologically desirable consequences, but in the more spiritual sense that our very affirmation helps to sustain the universal process. John Collier, in his book on American Indian cultures, *On the Gleaming Way*, quotes the Hopi belief that to be effective in sustaining the natural order,

> man must participate not merely by performing certain rites at prescribed intervals in certain ways; but he must also participate with his emotions and thoughts, by prayer and willing.... Hopi traditional philosophy therefore ascribes to man a purposive, creative role in the development of his will. The universe is not conceived as a sort of machine at the mercy of the mechanical law. Nor is it viewed as a sum total of hostile, competitive forces struggling for existence. It is by nature a harmonious, integrated system operating rhythmically according to the principles of justice, and in it the key role is played by man.[7]

Collier comments on the "intense romanticism" of the religion of the southwest Indians, which teaches that "the universe is a living being, and that the universe requires of man an inner concentration and a sustained action of desire and will, to the end that the universe itself may 'carry on'."[8]

I do not think that deep ecology (as Ecosophy T) sees human will as *central* to cosmic self-realization. To do so would be anthropocentrism in another form, to which deep ecology would have to be opposed. Nevertheless, I do think that deep ecology permits us to believe that we can enhance nature by helping to increase the cosmic *conatus*. This is not a matter of conservation at the physical level, though such conservation is, as we have seen, a commitment of the identification thesis. It is rather a matter of attitude, a spiritual matter, calling for an outright affirmation of nature, that may be expressed in an infinite number of possible ways—from the private revering of a flower in hand, through the devotional study of the natural sciences, to full-blown ritual celebrations in the style of the American Indians. Indeed, the affirmation of nature in its widest sense may be present in every thought and gesture, may animate every detail of our lives—leading to a life which is lived in the image of the cosmic law, viz., the conatus, the will-to-exist. Such a life may indeed strengthen the pulse of the cosmos, and this may be the deeper meaning of conservation: to cultivate the affirmation of life and to sustain it in one's heart for a lifetime. This is why I said, at the start, that the friends that nature needs—now and always—are life-affirming people.

My conclusion, then, is that deep ecology does enable us to deepen our sense of self, to an almost unimaginable degree, and this identification with the underlying, cosmic principle of life does enable us to confront the possibility of the extinction of life on Earth with equanimity, and in a life-affirming way. Cultivating this equanimity is perfectly consistent with our taking all reasonable steps to preserve life on Earth. Such equanimity in the face of ecocatastrophe no more obviates the need for conservation than the fact that one has the capacity to face one's own personal death in a positive, life-affirming way obviates the need for one to take all reasonable steps to preserve one's life. And the celebration of existence inspired by deep identification is not only important for personal sanity, but is, I have suggested, the deeper dimension of conservation itself.

Notes

1. The axioms of the deep ecology movement were initially set forth by Arne Naess in his now historic paper, "The Shallow and the Deep, Long-Range Ecology Movement," *Inquiry* 16 (1973): 95–100. They are elaborated by Bill Devall and George Sessions in *Deep Ecology: Living as if Nature Mattered* (Salt Lake City: Peregrine Smith, 1985).

2. See for example, Arne Naess, "Identification as a Source of Deep Ecological Attitudes," in Michael Tobias, ed., *Deep Ecology* (San Diego: Avant Books, 1985).

3. Spinoza, *Ethics*, trans. R.H.M. Elwes (New York: Dover Publications, 1951), part 3. prop. 7.

4. This justification occurs in my book on ecological metaphysics, *The Ecological Self*, 1991.

5. Even a Leibnizian monad is not strictly *identical with* the manifold of all other monads, even though the individual monad is wholly and logically constituted by its relations with the other monads. A monad may properly "identify with" the manifold of monads, inasmuch as the existence of the other monads is a condition for its self-realization.

6. The Arthurian myth of the wasteland is one example of the recurring archetype of the land which has lost the will-to-live, in which fertility declines to zero, and all beings turn away from life.

7. John Collier, *On the Gleaming Way* (Denver: Sage Books, 1949), pp. 101–102.

8. Ibid., p. 29.

Transpersonal Ecology and
the Varieties of Identification

Warwick Fox

Three Bases of Identification

How does one realize, in a this-worldly sense, as expansive a sense of self as possible? The transpersonal ecology answer is: through the process of identification. As Arne Naess says: "The ecological self of a person is that with which this person identifies. This key sentence (rather than definition) about the self, shifts the burden of clarification from the term 'self' to that of 'identification,' or rather 'process of identification.'"[1] How, then, does one proceed in realizing a way of being that sustains the widest and deepest possible identification? I suggest that there are three general kinds of bases for the experience of commonality that we refer to as identification; three general kinds of ways in which we may come to identify more widely and deeply. I refer to these bases of identification as *personal*, *ontological*, and *cosmological*.

Personally based identification refers to experiences of commonality with other entities that are brought about through personal involvement with these entities. This is the way in which most of us think of the process of identification most of the time. We generally tend to identify most with those entities with which we are often in contact (assuming our experiences of these entities are of a generally positive kind). This applies not only to concrete entities (e.g., the members of our family, our friends and more distant relations, our pets, our homes, our teddy bear or doll) but also to those more abstract kinds of entities with which we have considerable personal involvement (our football or basketball club, the individual members of which may change from year to year; our country). We experience these entities as part of "us," as part of our identity. An assault upon their integrity is an assault upon our integrity.

In contrast to personally based identification, ontologically and cosmologically based forms of identification are transpersonal in that they are not primarily a function of the personal contacts or relationships of this or that particular person. There is, of course, a sense in which *all* forms of identification beyond one's egoic, biographical, or personal sense of self can be described as *transpersonal*. However, the point here is that personally based identification is, as its name suggests, a far more personal—or, alternatively, a far less transpersonal—form of identification than either ontologically or cosmologically based identification, since it is a function of the personal contacts or relationships of this or that particular person, whereas, as we shall see below, the latter two forms of identification are not.

Ontologically based identification refers to experiences of commonality with all that is that are brought about through deep-seated realization of the fact *that* things are. (I am using the complex and variously employed term *ontology* in this context to refer to the fact of existence per se rather than to refer to the question of what the basic aspects of existence are or how the world is.) This is not a simple idea to communicate in words! Moreover, I do not intend to say very much about this idea since, in my view, it properly belongs to the realm of the training of consciousness (or perception) that is associated, for example, with Zen Buddhism, and those who engage in such training continually warn about the limits of language in attempting to communicate their experientially based insights. Martin Heidegger is a notable Western philosopher who does attempt to convey such insights in words, but then, although deeply rewarding, he is also notorious for the difficulty of his language. It is interesting to note in this connection, however, that upon reading a book by the Zen master D.T. Suzuki, Heidegger is reported to have said, "If I understand this man correctly, this is what I have been trying to say in all my writings."[2]

The basic idea that I am attempting to communicate by referring to ontologically based identification is that the fact—the utterly astonishing fact—that things *are* impresses itself upon some people in such a profound way that all that exists seems to stand out as foreground from a background of nonexistence, voidness, or emptiness—a background from which this foreground arises moment by moment. This sense of the specialness or privileged nature of all that exists means that "the environment" or "the world at large" is experienced not as a mere backdrop against which our privileged egos and those entities with which they are most concerned play themselves out, but rather as just as much an expression of the manifesting of Being (i.e., of existence per se) as we ourselves are. We have perhaps all experienced this state of being, this sense of commonality with all that is simply by virtue of the fact *that* it is, at certain moments. Things *are*! There is something rather than nothing! Amazing! If we draw upon this experience we can then gain some insight into why it is that people who experience the world in this way on a regular or semi-regular basis (typically as the result of arduous spiritual discipline) find themselves tending to experience a deep but impartial sense of identification with *all* existents. We can gain some insight into why such people find themselves spontaneously inclined "to be open for the Being [the sheer manifesting] of [particular] beings" and, hence, why, for them, "the best course of 'action' is to let beings be, to let them take care of themselves in accord with their own natures."[3]

For those who cannot see any logical connection between deep-seated realization of the fact that things *are* and the experience of deep-seated commonality with—and thus respect for—all that is, I can only reiterate that these remarks cannot and should not be analyzed through a logical lens. We are here in the realm of what Wittgenstein referred to as the mystical when he said, "It is not *how* things are in the world that is mystical, but *that* it exists."[4] If one seriously wishes to pursue the question of ontologically based identification then one must be prepared to undertake arduous practice of the kind that is associated with certain kinds of experientially based spiritual disciplines. Roger Walsh captures what is of central interest about these disciplines in this context by referring to them as *consciousness disciplines* in order to distinguish them "from the religious dogma, beliefs, and cosmologies to which most religious devotees adhere, and from the occult popularisms of both East and West".[5] Those who are not prepared to do this—that is, most of us—are no more in a position to dismiss the fruits of such practice than are people who would dismiss the fruits of scientific research without being prepared to undertake the training that is necessary to become a scientist or at least to understand the general features of scientific procedure.[6]

Cosmologically based identification refers to experiences of commonality with all that is that are brought about through deep-seated realization of the fact that we and all other entities are aspects of a single unfolding reality. This realization can be brought about through the empathic incorporation of any cosmology (i.e., any fairly comprehensive account of *how* the world is) that sees the world as a single unfolding process—as a "unity in process," to employ Theodore Roszak's splendid phrase.[7] This means that this realization can be brought about through the empathic incorporation of mythological, religious, speculative philosophical, or scientific cosmologies.[8] I am not meaning to assert by this that these various kinds of accounts of how the world is are equal in epistemological status, only that each is *capable* of provoking a deep-seated realization that we and all other entities are aspects of a single unfolding reality. Consider, for example, the world-views of certain indigenous peoples (e.g., of some Aboriginal North Americans), the philosophy of Taoism, or the philosophy of Spinoza.

For many people in the modern world the most viable—perhaps the only truly viable—source of cosmological ideas is science. Yet, despite this, there are many other people (including many who are formally trained in science or who simply have a general interest in science) who seem unable or unwilling to see science in a cosmological light. For them, science is all about prediction, manipulation, and control ("instrumental rationality") and cosmology is seen as something that belongs to mythology, religion, or speculative philosophy, or else as a highly specialized sub-discipline of physics that deals with the evolution and structure of the physical universe. But the anthropocentrically fueled idea that science is all about prediction, manipulation, and control is only half the story. As George Sessions says, "Modern science ... [has] turned out to be a two-edged sword."[9] The other side of science is its importance for understanding our place in the larger scheme of things (and it is scarcely necessary to add that this aspect has had profoundly *non*anthropocentric implications). This side of science is its cosmological aspect. Considered from this side, modern science can be seen as providing an account of creation that is the equal of any mythological, religious, or speculative philosophical account in terms of scale, grandeur, and richness of detail. More specifically, modern science is providing an increasingly detailed account of the physical and biological evolution of the universe that compels us to view reality as a single unfolding process.[10]

The most obvious feature of the physical and biological evolution of the universe as revealed by modern science is the fact that it has become increasingly differentiated over time. This applies not just at the level of biological evolution but also at the level of the physical evolution of the cosmos. If we think of this process of increasing differentiation over time diagrammatically then it is natural to depict it as a branching tree. Indeed, this is precisely the way in which evolutionary theorists think of biological evolution.[11] In general terms, ancestral species do not change into newer species; rather, newer species radiate out (branch away) from ancestral species, which can continue to exist alongside the newer species. This "budding off" process occurs when populations of a particular kind of organism become in any way reproductively isolated (e.g., through geographical divergence or through divergence in breeding seasons) and then undergo changes in their genetic composition, primarily as a result of natural selection, to the point where members of one population are no longer capable of interbreeding with members of the other population.[12] But it is not only phytogenetic development (the evolution of species) that must be depicted as a continually branching tree. The image of a branching tree is just as relevant to other forms of development that involve increasing differentiation over time, whether it be ontogenetic development (the evolution of individual organisms from a cell to maturity) or the evolution of the universe itself from *nothing* to its present state some fifteen billion years later.[13] As the science writer Stephen Young explains in a brief recent introduction to the importance of the tree metaphor in science generally: "Trees are indispensable to science. From physics to physiology, they serve as metaphors, expressing in a word details that would otherwise occupy a paragraph.... The theory of evolution is unthinkable without trees. Elsewhere within science, afforestation continues apace. If trees did not exist, scientists would have to invent them."[14]

Even if our present views on cosmological evolution (including phytogenetic and ontogenetic evolution) turn out to stand in need of modification in crucial respects, we still have every reason to believe that the particular views that supersede these views will be entirely in conformity with the far more general idea that all entities in the universe are aspects of a single unfolding reality that has become increasingly differentiated over time. The justification for such confidence lies not only in the fact that *all* the evidence that bears on this question across *all* scientific disciplines points in this general direction, but also in the fact that even the most radical scientific (i.e., empirically testable) challenges to our present scientific views also point in this general direction. What is at issue in scientifically framed debates about the evolution of the universe or the evolution of life is only the question of the *mechanisms* of evolution (i.e., the mechanisms that underlie the increasing differentiation of the universe over time), not the fact of evolution per se.

A good illustration of this general point is provided by Rupert Sheldrake's *hypothesis of formative causation*, which constitutes a fundamental challenge to our present understanding of the development of form in ontogenetic, phylogenetic, and cosmological evolution.[15] Sheldrake's hypothesis suggests that the form that every entity takes is shaped by, and in turn contributes to the shaping of, a formative field—Sheldrake calls it a *morphic field*—that is associated with that particular kind of entity. Thus, the physical forms of crystals, daisies, and elephants are supposed to be influenced by the morphic fields that have been built up by earlier examples of their own kind. Even the *behavior* of organisms is supposed to be influenced by the morphic fields that have been built up by the behaviors of earlier examples of their own kind. This suggests that people who have never learned Morse code, for example, ought to learn real Morse code faster or more accurately than a comparable group of people who are asked to learn a newly created version of Morse code. In Sheldrake's view this would be expected simply because many people have previously learned the real version of Morse code, and thereby contributed to the creation of a morphic field for the learning of that code, whereas no morphic field exists for the newly created code. As it happens, this experiment has been performed and the results support Sheldrake's hypothesis.[16]

Now if Sheldrake's fascinating but presently highly controversial hypothesis turns out to be supported by a wide range of experimental findings, then this would, I think, cause the biggest revolution in biology, and in the sciences generally, since Darwin's theory of evolution by means of natural selection. But the point in this context is this: even if Sheldrake's ideas were accepted, we would still find ourselves living in an evolutionary, "branching tree" universe because, as Sheldrake explains, the idea of formative causation does not reject Darwinian evolution but rather "greatly extends Darwin's conception of natural selection to include the natural selection of morphic fields."[17] Thus, even when we consider a challenge to mainstream science that is as broad and as profound in its implications as Sheldrake's, we find, as I have already stated, that what is at issue is still only the question of the *mechanisms* that underlie evolutionary processes, not the fact of evolution per se. Evolutionary development, in other words, is the great unifying theme of modern science.[18]

If we empathically incorporate (i.e., have a lived sense of) the evolutionary, "branching tree" cosmology offered by modern science then we can think of ourselves and all other presently existing entities as leaves on this tree—a tree that has developed from a single seed of energy and that has been growing for some fifteen billion years, becoming infinitely larger and infinitely more differentiated in the process. A deep-seated realization of this cosmologically based sense of commonality with all that is leads us to identify ourselves more and more with the entire tree rather than just with our leaf (our personal, biographical self), the leaves on our twig (our family), the leaves we are in close proximity to on other twigs (our friends), the leaves on our minor sub-branch (our community), the leaves on our major sub-branch (our cultural or ethnic grouping), the leaves on our branch (our species), and so on. At the limit, cosmologically based identification, like ontologically based identification, therefore leads to impartial identification with *all* particulars (all leaves on the tree).

Having said this, it must immediately be noted that, as with ontologically based identification, the fact that cosmologically based identification tends to be more *impartial* than personally based identification does not mean that it need be any less deeply felt. Consider the Californian poet Robinson Jeffers. For Jeffers, "This whole [the universe] is *in all its parts* so beautiful, and is felt by me to be so intensely in earnest, that I am *compelled* to love it" (emphases added).[19] Although Jeffers may represent a relatively extreme exemplar of cosmologically based identification, it should nevertheless be clear that this form of identification issues at least—perhaps even primarily? —in an orientation of steadfast (as opposed to fair-weather) friendliness. Steadfast friendliness manifests itself in terms of a clear and steady expression of positive interest, liking, warmth, goodwill, and trust; a steady predisposition to help or support; and, in the context of these attributes, a willingness to be firm and to criticize constructively where appropriate. Indeed, if a particular entity or life form imposes itself unduly upon other entities or life forms, an impartially based sense of identification may lead one to feel that one has no real choice but to *oppose*—even, in extreme cases, to terminate the existence of—the destructive or oppressive entity or life form. Even here, however, an impartially based sense of identification leads one to oppose destructive or oppressive entities or life forms in as educative, least disruptive, and least vindictive a way as possible.

Over time, steadfast friendliness often comes to be experienced by the recipient as a deep form of love precisely because it does not cling or cloy but rather gives the recipient "room to move," room to be themselves. It may be of particular interest to add here that Arne Naess seems to me to be an exemplar of steadfast friendliness—and of course I am not only talking here about his relationship with me over the years but of his orientation toward the world in general. It is also interesting to note that Naess has himself written a paper on the importance of the concept of friendship in Spinoza's thinking in which he notes that "the intellectual sobriety of Spinoza favors *friendship rather than worship*" and that, for Spinoza, "friendship is the basic social relation" between members of a free society.[20] Naess concludes this paper by explicitly linking the theme of friendship in Spinoza's philosophy with "the ecological concept of symbiosis as opposed to cutthroat competition." "Both in Spinoza and in the thinking of the field ecologist," says Naess, "there is respect for an extreme diversity of beings capable of living together in an intricate web of relations."[21]

Notwithstanding the eloquent testimonies to cosmologically based identification that have been offered by Spinoza, Gandhi, Jeffers, Naess, and many others (even Einstein, for example), many people find it difficult to think of identification in anything other than personally based terms. For these people, cosmologically based identification approximates to something like going out, encountering every entity in the universe (or at least, on the planet) on a one-to-one basis, and coming to identify with each entity on the basis of that contact. But this simply represents an example of personally based identification that has been blown up into universal (or global) proportions. In contrast, cosmologically based identification means having a lived sense of an overall scheme of things such that one comes to feel a sense of commonality with all other entities (whether one happens to encounter them personally or not) in much the same way as, for example, leaves on the same tree would feel a sense of commonality with each and every other leaf if, say, we assumed that these leaves were all conscious and had a deep-seated realization of the fact that they all belonged to the same tree. In summary, then, personally based identification proceeds from the person—and those entities that are psychologically, and often physically, closest to the person—and works outward to a sense of commonality with other entities. In contrast, cosmologically based identification proceeds from a sense of the cosmos (such as that provided by the image of the tree of life) and works inward to each particular individual's sense of commonality with other entities. In vectorial terms, this contrast in approaches means that we can think of personally based identification as an "inside-out" approach and cosmologically based identification as an "outside-in" approach.

One may gain or seek to cultivate a cosmologically based sense of identification in a wide variety of ways. Even if we exclude mythological, religious, and speculative philosophical cosmologies and restrict ourselves to the cosmology of modern science, these ways of coming to embody a cosmologically based sense of identification can range from approaches such as the ritualized experientially based work being developed by John Seed and Joanna Macy under the title "Council of All Beings"[22] [see the articles in this Anthology]; to participation in theoretical scientific work (a number of the very best scientists have had a profound sense of cosmologically based identification); to more practically oriented involvement in natural history (many naturalists and field ecologists, for example, effectively come to experience themselves as leaves on the tree of life and seek to defend the unfolding of the tree in all its aspects as best they can); to simply developing a deeper personal interest in the scientific world model and in natural history along with one's other interests.

Identification, Delusion, and Enlightenment

Having now identified three general forms of identification, it is important to consider the question of the role that these forms of identification play in *transpersonal ecology* [italics added]. To begin with, it is important to note that transpersonal ecologists often discuss identification in terms of personally based identification. This, of course, is only to be expected: most of us are so familiar with this form of identification and are so used to thinking of the process of identification in this way that it is difficult *not* to discuss identification in terms of personally based identification. But, having said that, the interesting point to note here is that the *theoretical* emphases in transpersonal ecology lie squarely with the two transpersonal forms of identification (i.e., ontologically and cosmologically based identification). More specifically, although some Heideggerian and Zen-influenced transpersonal ecologists have tended to emphasize an ontological basis for wider and deeper identification, most of the main writers on transpersonal ecology have followed Naess in emphasizing a cosmological basis. For Naess, identification has a cosmological basis in that it follows (psychologically) *from* the realization that "life is fundamentally one."[23]

The fact that Naess emphasizes a cosmologically based approach to identification is clear not only from his writings but also from the nature of the sources that inspired his central concept of Self-realization. The inspiration for this concept derives from Gandhi and Spinoza, both of whom explicated their views within the context of a monistic metaphysics, that is, within the context of a cosmology that emphasized the fundamental unity of existence. Gandhi was committed to Advaita Vedanta (i.e., monistic or, more literally, nondual Hinduism), to the belief that all life comes from "the one universal source, call it Allah, God or Parmeshwara."[24] He expressed this belief by conceiving of all entities as drops in the ocean of life: "The ocean is composed of drops of water; each drop is an entity and yet it is a part of the whole; 'the one and the many.' In this ocean of life, we are little drops. My doctrine means that I must identify myself with life, with everything that lives, that I must share the majesty of life in the presence of God. The sum-total of this life is God."[25]

Spinoza developed a philosophy that conceived of all entities as *modes* of a single *substance*—or what might today be thought of as expressions of a single energy.[26] As Roger Scruton notes in a recent study of Spinoza: "There is a modern equivalent of Spinoza's monism in the view that all transformations in the world are transformations of a single stuff—matter for the Newtonians, energy for the followers of Planck and Einstein."[27] For Spinoza, the highest end to which humans could aspire consists in "knowledge of the union existing between the mind and the whole of nature."[28] Thus, humans (one particular kind of mode) realize the truth of existence, or attain self-realization, when they realize that they arise out of and so are united with "the whole of nature," the single substance (or energy) that constitutes all modes of existence.

Despite the fact that transpersonal ecologists have generally emphasized cosmologically based identification more than ontologically based identification, I do not think that there is any particular *theoretical* reason for preferring one of these approaches to the other in transpersonal ecology. There may be a *practical* reason for this emphasis, however, in that it would seem to be much easier to communicate and inspire a cosmologically based sense of identification with all that is rather than an ontologically based sense of identification with all that is. This should be apparent from my previous comments to the effect that, beyond a certain point (namely, drawing attention to the fact *that* the world is), one simply cannot *say* very much about ontologically based identification. One can only direct those who are interested in deep-seated realization of the fact of Being (the fact of existence per se) to the consciousness disciplines—and no one should doubt that these disciplines *are* disciplines, and arduous ones at that. Anyone who mistakes this pointing *to* the consciousness disciplines for insight into the fact of Being itself is, as the Zenists say, mistaking a finger that is pointing at the moon for the moon itself. In contrast, the fact that cosmologies (i.e., general accounts of *how* the world is) are formulated in both words and images means that cosmologically based identification can readily be inspired through symbolic communication. These forms of symbolic communication may range from the communication of scientific, speculative philosophical, religious, and mythological views to the communication of vivid visual images such as mandalas and the kinds of images that have been presented here in which entities have been conceived of as leaves on a tree (as I have urged), drops in the ocean (Gandhi), or, more abstractly, as modes of a single substance (Spinoza).

I want to turn now from the emphasis that is placed in transpersonal ecology on ontologically and, especially, cosmologically based forms of identification and consider the relative lack of emphasis that is placed on personally based identification. In contrast to the other two forms of identification, there is a fundamental theoretical reason why transpersonal ecologists do not emphasize personally based identification. Specifically, the fact that personally based identification refers to experiences of commonality with other entities that are brought about through personal involvement with these entities means that this form of identification *inevitably* leads one to identify most with those entities with which one is most involved. That is, one tends to identify with my self first, followed by *my* family, then *my* friends and more distant relations, *my* cultural or ethnic grouping next, *my* species, and so on—more or less what the sociobiologists say we are genetically predisposed to do. The problem with this is that, while extending love, care, and friendship to one's nearest and dearest is laudable in and of itself, the *other* side of emphasizing a purely personal basis for identification is that its practical upshot (*my* self first, *my* family and friends next, *my* cultural or ethnic grouping next, *my* species next, and so on) would seem to have far more to do with the *cause* of possessiveness, greed, exploitation, war, and ecological destruction than with the solution to these seemingly intractable problems.

I can hardly stress the importance of this last point enough. Personally based identification can slip so easily—and imperceptibly—into attachment and proprietorship. Anybody who doubts that personally based identification is a potentially treacherous basis for identification need only reflect on the way in which romantic love between two people—a paradigmatic example of intense personally based identification—can sometimes collapse into acrimonious divorce; or the truth in the old adage about family fights often being the worst fights; or the fact that we will do things to "others" (or allow things to be done to "them") that we would never do (or allow to happen) to "one of us." Yet again and again ecophilosophical discussants are all too prepared to extol the virtues of what effectively amounts to personally based identification while simply ignoring the possessive, greedy, exploitative, warmongering, and ecologically destructive drawbacks that can also attend this particular basis of identification.

It has been important to go beyond transpersonal ecology's concern with identification in general and to consider the particular kinds of identification that are emphasized in transpersonal ecology. However, having done this, it is also important to note that the emphasis that transpersonal ecologists place on ontologically and, particularly, cosmologically based identification is just that—an emphasis, not an outright *opposition* to personally based identification. Thus, for example, when Naess says that "a rich variety of acceptable motives can be formulated for being more reluctant to injure or kill a living being of kind A than a being of kind B," he makes it clear that "felt nearness" (i.e., personally based identification) is an important example of such a motive.[29] I have simply emphasized the negative aspects of personally based identification here in order to show why transpersonal ecologists choose to emphasize transpersonally based forms of identification (and, in the process, as a corrective to the one-sided way in which personally based identification is often presented by other ecophilosophical discussants). But transpersonal ecologists still recognize that personally based identification is an inescapable aspect of living and that it plays a fundamental role in human development.

The upshot, then, is that transpersonal ecologists simply want to put personally based identification in what they see as its appropriate place. For transpersonal ecologists, ontologically and cosmologically based identification are seen as providing a *context* for personally based identification. This context serves as a corrective to the partiality and problems of attachment that are associated with personally based identification. When personally based identification is set within the context of ontologically and cosmologically based forms of identification (i.e., within the context of forms of identification that tend to promote impartial identification with *all* entities) then it is expressed in terms of a person being, as Naess says, more *reluctant* to interfere with the unfolding of A than B in those situations where a choice is unavoidable if the person is to satisfy nontrivial needs of their own. However, considered in the absence of the overarching context provided by ontologically and cosmologically based identification, personally based identification is expressed in terms of a person having no desire to harm A in any way (say, their child) but having few or no qualms about interfering with—or standing by while others interfere with—the unfolding of B (where B is an entity of any kind—plant, animal, river, forest—with which the person has no particular personal involvement). And this, of course, can apply even when the reasons for interfering with B are relatively trivial.

The significance of these considerations should be clear: although the *positive* aspects of personally based identification are praiseworthy and fundamental to human development, the *negative* aspects that go with exclusive or primary reliance upon this form of identification (my self first, my family and friends next, and so on) are costing us the Earth. They underlie the egoisms, attachments, and exclusivities that find personal, corporate, national, and international expression in possessiveness, greed, exploitation, war, and ecocide. As an antidote to these poisons, transpersonal ecologists emphasize the importance of setting personally based identification firmly within the context of ontologically and cosmologically based identification—forms of identification that lead to impartial identification with all entities. In terms of politics and lifestyles, the latter, transpersonal forms of identification are expressed in actions that tend to promote the freedom of all entities to unfold in their own ways; in other words, actions that tend to promote symbiosis. (Indeed, Naess says at one point that "if I had to give up the term ['Self-realization!'] fearing its inevitable misunderstanding, I would use the term 'symbiosis.'"[30]) Actions of this kind include not only actions that consist in "treading lightly" upon the earth (i.e., lifestyles of voluntary simplicity) but also actions that respectfully but resolutely attempt to alter the views and behavior of those who persist in the delusion that self-realization lies in the direction of dominating the Earth and the myriad entities with which we coexist.

> *That the self advances and confirms the*
> *myriad things is called delusion.*
> *That the myriad things advance and confirm*
> *the self is enlightenment.*[31]

Notes

1. Arne Naess, "Self-realization: An Ecological Approach to Being in the World," *The Trumpeter* 4(3) (1987): 35–42, p. 35.

2. Quoted in William Barrett, "Zen for the West," in *Zen Buddhism: Selected Writings of D.T. Suzuki*, ed. William Barrett (Garden City, NY: Doubleday/Anchor Books, 1956), p. xi. There is a whole literature on the similarities between Heidegger's thought and Eastern thought, especially Zen. For a guide to much of this literature, see the papers and books listed at note 3 in Michael Zimmerman, "Heidegger and Heraclitus on Spiritual Practice," *Philosophy Today* 27 (1983): 87–103. Special mention should be made here of Zimmerman's own book on Heidegger entitled *Eclipse of the Self: The Development of Heidegger's Concept of Authenticity* (Athens: Ohio University Press, 1981), which explores the relationship between Heidegger's thought and Zen in its final section (pp. 255–76). In addition to the papers and books cited by Zimmerman in "Heidegger and Heraclitus," see the following inspirational papers by Hwa Jol Jung: "The Ecological Crisis: A Philosophic Perspective, East and West," *Bucknell Review* 20 (1972): 25–44; and "The Paradox of Man and Nature: Reflections on Man's Ecological Predicament," *The Centennial Review* 18 (1974): 1–28.

3. Michael Zimmerman, "Toward a Heideggerean Ethos for Radical Environmentalism," *Environmental Ethics* 5 (1983): 99–131, pp. 102 and 115.

4. Ludwig Wittgenstein, *Tractatus Logico-Philosophicus*, trans. D.F. Pears and B.F. McGuiness (London: Routledge and Kegan Paul, 1961), proposition 6.44.

5. Roger Walsh, "The Consciousness Disciplines and the Behavioral Sciences: Questions of Comparison and Assessment," *American Journal of Psychiatry* 137 (1980): 663–73, p. 663.

6. On this general point, see Ken Wilber's insightful essays "Eye to Eye" and "The Problem of Proof," which constitute the first two chapters of his book *Eye to Eye: The Quest for the New Paradigm* (Garden City, NY: Anchor Books, 1983).

7. Theodore Roszak, *Where the Wasteland Ends: Politics and Transcendence in Postindustrial Society* (London: Faber and Faber, 1973), p. 400.

8. On the general question of the empathic incorporation of cos-

mologies or "world models," see Alex Comfort, *Reality and Empathy: Physics, Mind, and Science in the 21st Century* (Albany: State University of New York Press, 1984). By *empathy*, Comfort means an "incorporation going beyond intellectual assent" (p. xviii). See also Stephen Toulmin, *The Return to Cosmology: Postmodern Science and the Theology of Nature* (Berkeley: University of California Press, 1982), esp. the final chapter in which Toulmin explicitly links the cultivation of a cosmological sense of things—or what I am referring to as cosmologically based identification—with the development of "a genuine piety ... toward creatures of other kinds: a piety that goes beyond the consideration of their usefulness to Humanity as instruments for the fulfillment of human ends" (p. 272).

9. George Sessions, "Ecocentrism and the Greens: Deep Ecology and the Environmental Task," *The Trumpeter* 5 (1988): 65–69, p. 67.

10. One could drown in the number of semi-popular and more technical books that could be cited at this point! A gentle approach might be more effective; thus, for a highly readable, comprehensive, *single* volume overview of the scientific view of the world, see Isaac Asimov's exemplary guide *Asimov's New Guide to Science*, rev. ed. (Harmondsworth, Middlesex: Penguin Books, 1987). For an excellent systems-oriented overview of the scientific view of the world, see Ervin Laszlo, *Evolution: The Grand Synthesis* (Boston: Shambhala, 1987).

11. See, for example, Richard Dawkins, *The Blind Watchmaker* (London: Penguin Books, 1988), esp. ch. 10: "The One True Tree of Life."

12. See Mark Ridley, *The Problems of Evolution* (Oxford: Oxford University Press, 1985), ch. 8: "How Can One Species Split into Two?"

13. For overviews of recent work on the origins of the physical cosmos, see Paul Davies, *God and the New Physics* (Harmondsworth, Middlesex: Penguin Books, 1984); Paul Davies, *Superforce: The Search for a Grand Unified Theory of Nature* (London: Unwin Paperbacks, 1985); John Gribbin, *In Search of the Big Bang: Quantum Physics and Cosmology* (London: Corgi Books, 1987); Alan H. Guth and Paul J. Steinhardt, "The Inflationary Universe," *Scientific American* (May 1984): 90–102; Stephen W. Hawking, *A Brief History of Time: From The Big Bang to Black Holes* (New York: Bantam Books, 1988); and Heinz R. Pagels, *Perfect Symmetry: The Search for the Beginning of Time* (New York: Bantam Books, 1986).

14. Stephen Young, "Root and Branch in the Groves of Academe,"

New Scientist (23–30 December 1989): 58–61, at pp. 58 and 61.

15. Rupert Sheldrake, *A New Science of Life: The Hypothesis of Formative Causation*, 2nd ed. (London: Paladin/Grafton Books, 1987); and Rupert Sheldrake, *The Presence of the Past: Morphic Resonance and the Habits of Nature* (New York: Vintage Books, 1989).

16. Arden Mahlberg, "Evidence of Collective Memory: A Test of Sheldrake's Theory," *Journal of Analytical Psychology* 32 (1987): 23–34. Sheldrake's two books provide overviews of the experimental work that has been performed to test his hypothesis thus far. The appendix in the second edition of *A New Science of Life* also contains an overview of the controversy that greeted the original publication of his hypothesis.

17. Sheldrake, *The Presence of the Past*, p. 294.

18. See Ervin Laszlo, *Evolution: The Grand Synthesis*.

19. Quoted in Bill Devall and George Sessions, *Deep Ecology: Living as if Nature Mattered* (Salt Lake City: Peregrine Smith Books, 1985), p. 101.

20. Arne Naess, "Friendship, Strength of Emotion, and Freedom," in *Spinoza Herdacht 1677–1977* (no further details), pp. 11–19, at p. 13.

21. Ibid., p. 19.

22. See John Seed, Joanna Macy, Pat Fleming, and Arne Naess, *Thinking Like a Mountain: Towards a Council of All Beings* (Santa Cruz: New Society Publishers, 1988).

23. See, for example, Arne Naess, *Ecology, Community and Lifestyle: Outline of an Ecosophy*, trans. and rev. David Rothenberg (Cambridge: Cambridge University Press, 1989), ch. 7: "Ecosophy T: Unity and Diversity of Life" (the "Life is fundamentally one" quotation is from p. 166 of this chapter).

24. Quoted in Ramashray Roy's illuminating study in Gandhian thought: *Self and Society: A Study in Gandhian Thought* (New Delhi: Sage Publications, 1984), p. 73. Having met Gandhi as well as studied his thought, Ronald Duncan observes that "Gandhi's whole philosophy was based on the oneness or the wholeness of life" (Mahatma Gandhi, *The Writings of Gandhi: A Selection*, ed. Ronald Duncan [London: Fontana/Collins, 1983], p. 29).

25. Quoted in Roy, *Self and Society*, p. 103. The chapter of Roy's book from which this quotation is taken is entitled "The Drop and the Ocean" and represents an extended discussion of the point that Gandhi

makes in this quotation. It is interesting to note that Roy explicates Gandhi's metaphysics in this chapter in an extremely Naessian way (despite the fact that he does not appear to be familiar with Naess' work). Thus, Roy explains that, for Gandhi, "self-realization ... is manifested in an endeavor to identify with the surging sea of life without—life that, in sum-total, is God. In seeking such an identity the self extends beyond its physical boundaries ... [so that] society is not conceived as something out there but as an extended self" (p. 106).

26. Baruch Spinoza, *The Ethics and Selected Letters*, trans. Samuel Shirley and ed. Seymour Feldman (Indianapolis: Hackett Publishing Company, 1982).

27. Roger Scruton, *Spinoza*, Past Masters series (Oxford: Oxford University Press, 1986), p. 51.

28. From Spinoza's *On the Improvement of the Understanding* in *The Chief Works of Benedict de Spinoza*, trans. R.H.M. Elwes (New York: Dover, 1955), p. 6.

29. Arne Naess, "Intuition, Intrinsic Value, and Deep Ecology," *The Ecologist* 14 (1984): 201–203, 202. Naess fully accepts that "any realistic praxis necessitates some killing, exploitation, and suppression" ("The Shallow and the Deep, Long-Range Ecology Movement: A Summary," *Inquiry* 16 (1973): 95–100, p. 95.

30. Arne Naess, "The Deep Ecological Movement: Some Philosophical Aspects," *Philosophical Inquiry* 8 (1986): 10–31, p. 28.

31. Dogen Zenji (1200–1253), quoted in Robert Aitken, "Gandhi, Dogen and Deep Ecology," appendix C in Devall and Sessions, *Deep Ecology*, p. 232; from *The Way of Everyday Life: Zen Master Dogen's Genjokoan with Commentary*, trans. Hakuyu Taizan Maezumi (Los Angeles: Center Publications, 1978), n.p.

12

A Platform of Deep Ecology

David Rothenberg

Introduction

The term 'deep ecology' was introduced in the early 1970s by a group of Norwegian environmentalists, particularly the philosopher Arne Naess (1973—first English publication of this definition). The term is meant to characterize a way of thinking about environmental problems that attacks them from the roots, i.e., the way they can be seen as symptoms of the deepest ills of our present society. This is to be contrasted with 'shallow ecology' —treating merely the symptoms themselves, not the causes, through technological fixes such as pollution-control devices, regulations upon industry, etc.

The adherents to deep ecology movement usually do not criticize the latter means as being unnecessary, but rather imply that in themselves they are not sufficient. Such new technologies and reforms in our current system are clearly easier to implement than any fundamental, general changes. Unfortunately, the 'shallow' solutions also stay quite clear of the heart of the problem, and thus will eventually fail if not seen as part of a much larger and more difficult type of change. Although it is hard to reckon just what changes are necessary in our society, it is even harder to know just what changes of this albeit idealistic kind are possible, and then how to implement them or set the stage for them to happen of their own accord.

Any deeper change will take a great deal of time, and many people feel that, in environmental matters, there is little such time available to play with. But steps can be taken. To attempt to unify the intentions of the growing number of people around the globe who are concerned with approaching ecological problems in a more mature reflective manner, Arne Naess and California philosopher George Sessions set down a series of points for a preliminary 'platform' in 1984. These were supposed to be a series of standpoints that supporters of a deep ecological view [movement] would agree upon in general, proceeding to elaborate their own specific variants in different yet compatible ways.

This 'platform' has been published in several places (Devall and Sessions, 1985; Naess, 1986; Reed and Rothenberg, 1987), and generated a fair amount of critical response (Watson, 1983; 1985; Fox, 1984; Sylvan, 1985). Yet there seem to be enough people around the globe in support of these ideas to speak of a deep ecology movement in intellectual circles as well as among activists (both Australia's Rainforest Information Centre and USA's Earth First! group cite deep ecology as an inspiration and guiding philosophy). But the connection to realistic policy change has yet to be developed, and this is where work should be concentrated in the future.

A platform is supposed to be something that will incite agreement. So could others sit down and derive platforms of an essentially similar nature to the Naess/Sessions attempt? As an exercise, Arne Naess challenged me to come up with such a platform that would be accessible to a wider audience than the one for which he had written. The primary purpose is to present some fundamental ideas in outline form but also to inspire discussion; the kind of discussion that environmental thinking and planning sorely needs.

As a result, its language is not entirely that of an academic article. To communicate unfamiliar ideas it is sometimes necessary to experiment in language. Some might label this as a 'manifesto', but it is nothing fixed or inflexible. Many people should be able to feel 'at home' in this movement—it should encourage some to rethink and change what they have been doing and living, and others to realize they have been a part of it all along.

A Platform of Deep Ecology

Deep ecology is a way of rethinking the relationship of humanity and/in nature that has critical implications for envisioning ideal societies, preferred ways of life, and the development of practical strategies for approaching these ideals.

There are three clear purposes to this rethinking:

(1) It can provide a firm philosophical grounding for activism.

(2) It can encourage decision-makers to connect philosophical and religious assertion with concrete policy.

(3) It can be used to get as many people as possible to think about themselves and nature in a new way.

Seven basic points of deep ecology follow in outline form. They are derived from many sources, and are general enough to be interpreted in a variety of ways. They are meant to serve as a set of core values, a platform to guide discussion and action. They are neither the roots nor the branches of the tree of deep ecology, but the trunk itself. The central metaphor of the platform of deep ecology being like the trunk of a tree, distinguished from its branches and its roots, has to do with the notion of *derivation*, both logical and historical. Historically, one can ask the question "where did the ideas within deep ecology come from?" and then one can find many *roots* of support within various religious, literary, and speculative traditions (the best bibliography on deep ecology is Sessions, 1981). Logical roots (as described by Naess in Devall and Sessions, 1985, p. 226) are one's own personal religious or philosophical presuppositions from which the general and simple statements of the platform are derived within one's own thinking, e.g., "How do I justify my own support of these principles?" For both types of derivation, the *branches* of the tree are the various ways the principles are elaborated and applied in practice. Some of these are suggested in the conclusion to the platform printed here.

If something appears vague, do not ask for greater precision, but think to yourself, "Where does this value seem to come from for me? What are my interpretations?"

I. LIFE!	There is intrinsic value in all life.
II. NATURE!	Diversity, symbiosis, and thus complexity explain the life of nature itself.
III. HUMAN IN/AND NATURE!	Humanity is a part of nature, but our potential of power means that our responsibility towards the Earth is greater than that of any other species.
IV. NO FALSE DISTANCE!	We feel estranged from the Earth because we have imposed complication upon the complexity of nature.
V. OUTSIDE CHANGE!	On the outside, we should change the basic structures of our society and the policies which guide them.
VI. INSIDE CHANGE!	On the inside, we should seek quality of life rather than higher standard of living, self-realization rather than material wealth.
VII. SPREAD OF IDEAS!	New kinds of communication should be found that encourage greater identification with nature. Only then will we see our part in it again.
ACTION!	CONCLUSION: Those who accept the above points have an obligation to try to implement the necessary changes.

Seven Basic Points

I. *There is intrinsic value in all life*

In addition to all organic processes, the term *life* is here expanded to cover the physical life of an ecosystem, the interaction of ecosystems, and the inherent life of the Earth as a whole. The expansion of the term 'life' in this way requires the use of relational concepts to define and shape an image of the world.

II. *Diversity, symbiosis, and thus complexity explain the life of nature itself*

Diversity refers to the many different kinds of individuals, species, and ecosystems which the whole of nature contains. Diversity in itself implies the idea of 'many'. But in nature there is not a multiplicity of detached entities, but things bound to each other through threads of *symbiosis*—the dependence upon interaction with others for survival. A thing does not exist without its relation to other things.

Symbiosis with diversity together frame the *complexity* of nature— a vast world of relationships, connections, and possibilities that show how the many can be seen as one. There is intrinsic value in this crystal web of complexity. Our species is but one strand in this web.

The relationship between the concepts of complexity, diversity, and symbiosis could be considered the core of deep ecology—where three ideas from the science of ecology are applied in a philosophical frame. When anything is placed into a new context, debate ensues. I would say that complexity is the single word for the virtue and perfection of nature that we are slowly discovering more about, leading to an increase in awe and wonder. Diversity together with symbiosis are the particular qualifying factors that define and make possible this complexity. For Naess' views on the subject, see the interview with him in Reed and Rothenberg (1987).

III. *Humanity is a part of nature, but our potential of power means that our responsibility toward nature is greater than that of any other species*

Individuals of all species have a natural tendency to explore their environment and simultaneously fill and create an ecological niche. The defining of our niche has involved a greater alteration of nature than that of any other species. When this imposed change is carried out on a massive scale, it serves to detach us from the Earth, and then furthers neither our own survival nor the well-being of the planet itself. Yet the power of our species is more than power to destroy; it is also a potential for understanding.

IV. *We feel estranged from the Earth because we have imposed complication upon the complexity of nature*
There is a marked difference between the unified diversity of nature and the apparent manifold possibilities our present society offers us. We have many choices, but so many of them seem to make no sense at all.

We have learned to think of ourselves and our world as machines. A machine has many parts, but no internal intelligence or sense to guide it. The parts are not aware of their purpose and functions. One can usually find spare parts for a broken machine. Not so for the Earth.

Thinking *like* machines is only one step from trying to live *as* machines. A synthetic lifestyle alienates us while hiding the explicit reasons for alienation. It leads to economic problems through class separation, philosophical problems through exclusive thought in terms of either/or, yes/no. It leads to ecological problems, leading us to ask, "Must we choose between human *or* nature?" This type of thinking and living is *complication*, opposed to and apart from the *complexity* of nature.

For the solution of these deeply rooted problems, our dualistic thoughts and actions must be changed. One cannot change without the other, but we can identify two directions in which change can go:

(1) The pragmatic goal of change should be a respectful way of living on Earth,

(2) The ideal-centered goal should be a recognition of our own unfolding as a part of nature.

V. On the outside we should change the basic structures of our society and the policies which guide them
The idea of *growth* is to be redefined so that it refers to the increase of our understanding and experience of nature. It is clear that the capacity to reflect and interpret the natural complexity of the Earth is the basis of our potential to understand. Upon this principle we may identify three broad areas of structure and policy change:

(a) *Economics*: Economic needs are the factual needs of survival. Only living beings can have such needs, not organizations or corporations—these can only be instrumental. Present economic ideology tends to value that which can be put on the market: material things and the flow of goods and services. Industries try to create the needs for products if these needs do not legitimately exist. Economics by itself cannot deal with uniqueness and irreversibility.

We should instead identify clearly the essential needs of humans and those of other species, and develop ways of realizing them.

(b) *Technology and cultures*: The statement and realization of needs vary in different cultures. Cultures must be seen as dynamic patterns, flows of change based on enduring values identified through history. The continuance of cultural diversity today demands techniques that advance the basic aims of each culture. The tools appropriate to each particular situation are what we shall define as *appropriate* technology. And appropriate technology should be seen as the most *advanced* technology. Here the words appropriate and advanced are contrasted in reference to technology. There is not one kind of advanced technology, but there should be many, to encourage a diversity of cultures and ways of solving basic problems. There should be different kinds of technology, appropriate to the variety of situations around the world. People should not be encouraged to accept the same solutions everywhere.

Means cannot be separated from ends in the definition of culture, and the same for technology. No culture should impose anything blindly upon another.

(c) *Society and politics*: Ours is a world largely controlled by massive organizations. It is unrealistic to assume that we could function without them. But to promote growth we must encourage more local, participatory structures, based on the principles of *self-reliance*: Local autonomy does not imply isolation. Decentralization does not suggest a lack of cooperation and meeting. The movement towards smaller, more egalitarian and less hierarchical organizations should be attempted within the dominant system: otherwise the system itself will not change.

VI. *On the inside, we must seek quality of life rather than higher standard of living, self-realization rather than material wealth*
Measuring quantities in our lives is easy; measuring quality is not. Statistics are easy to collect, but they alone can never be the basis for decisions. Underlying values must be there. So it is essential to set forth principles of life quality before critiques and paths for change are mapped out.

Self-realization is a process that connects the individual to the larger world. It expresses the full unfolding of the possibilities open to any person in society and nature. As many great religions have taught us, no one's realization can come about without that of all: so compassion and altruism must be the foundation of any life that is truly to be one of quality.

This is certainly not a new idea (see Naess, 1974; 1979). Self-realization brings the many into one. It points to a type of understanding that is easy to feel, but quite difficult to describe. Better description will only be possible if we find new ways of sharing and exchanging our thoughts and ideas.

VII. *New kinds of communication should be found that encourage greater identification with nature. Only then will we see our part in it again*
There has been a tendency to illuminate nature in a mechanical way. Ecologists have compared the workings of the tropical rainforest to the division of labor in New York City. This is inadequate—the city has conditions of poverty and injustice which the rainforest does not share. The pristine ecosystem is far more perfect, and there is much we can learn from it. We should try to create and explain our organizations with structure and language *based in nature*.

Science can help us to do this, as the search for basic laws brings us closer to natural entities' processes that first appeared distant. Art, in its widest sense, including all pursuits of the imagination, can be encouraged to follow nature in its manner of operation. There are many ways the richness of nature can be made clearer and more relevant to all. And the more we learn of nature, the more we see that we cannot accept as inevitable our present and dangerously imperfect world. We cannot lie complacent.

Conclusion

Those who accept the above points have an obligation to try to implement the necessary changes

We can direct whatever abilities we possess towards change, both immediate action and the gradual clarification of long-term goals. Four broad ways of involving oneself in change can be identified:

(1) Living according to ecological ideals of self-reliance as an individual or in a small group. To do this at present requires some detachment from the dominant system.

(2) Encouraging compromise between the present and the ideal: a mix of centralized *and* local technologies and institutions, providing a realistic path of transition to a society that could exist in harmony with these ideals.

(3) Trying to change the system directly. This means talking to people:

    (a) the 'experts' and decision makers.

    (b) the rest of us.

Different languages and methods may be appropriate for each.

(4) Artistic and philosophical reflection on the closeness of humans and nature for its own sake.

There is no point in degrading the efforts of those working at different approaches. As an individual working in the ecological movement, one must define one's own place. It is important to concentrate only upon tasks which can be mastered: those upon which one's labor will not be wasted. The diversity of many methods and the symbiosis between them work together to enact a change greater than the sum of its parts.

As we do not live in an ideal world, we require ideals to guide our actions. The points of this platform should be interpreted in ways particular to our own situations. Finding oneself only begins by looking inward, as we quickly discover that any deep search soon leads outward, seeing ourselves only in terms of relationships of all kinds, identifying and discovering connections never before known.

"Supported by many roots, the tree grows tall and straight, but out of it sprout many branches and countless leaves that flutter in the wind."

Final Comments

How does one make use of such a platform of deep ecology? It presents only general statements of a type it is difficult for many to dispute. And many people find they agree in theory with all or most of the positions presented but still lead lives which indirectly support a kind of society in diametrical opposition to these principles.

Is this just a sign of the clearly hypocritical nature of our times? It has been said that the comparison between one's ideals and the way one acts in practice is the truest measure of one's quality of life. So perhaps most of us are today unable to live in a very contented or 'qualitatively' sufficient way.

But even once the problem is recognized what to do about it? Here is presented a series of slogans—perhaps just phrases to carry around and repeat to oneself or others. Yet there is a way forward: concentrate on the parts of the platform that deal with the foundation of the ideas. If one takes the notion of realization and expansion of the Self (as in Naess' Ecosophy T) seriously it involves considering more and more of nature and one's environment as an essential part of one's identity. Then one's view as to what is important and necessary in nature changes quite a bit; we feel more connected to the surroundings and thus feel more for the surroundings.

It is impossible to tell someone *how* to arrive at such a feeling, other than to suggest paths, conclusions, or suppositions that might be of some help or inspiration. Communicating with generalities has advantages and disadvantages. If a statement is general and sweeping, many can feel 'at home' with it, find it relevant and interpret it in their own ways. But clarity of intention is sacrificed: people are challenged, drawn in, and may react, but have little idea of what precisely is implied or intended.

On the other hand, speaking too specifically, i.e., "we must live in decentralized small communities," seems to imply a cold fixedness of a future where we do not have the opportunity to explore and shape things. Fewer people will be interested if they feel they are being told what to do. And there is, beneath all its qualifying statements, a certain absolutist character to this platform.

Yet we must move between and among these extremes. This platform has also been written by one who has heard many slogans of the ecology movement over and over again, in the end not knowing exactly what to do with them. I have tried to place such slogans into a kind of order through which one can pass logically, and conclude that whoever you are, there is something you can do towards insuring the continued and sustainable survival of life on this planet, and that this should be done!

References

Devall B. & Sessions, G. (1985) *Deep Ecology: Living as if Nature Mattered*. Peregrine Smith Books, Salt Lake City.

Fox, W. (1984) Deep Ecology: A New Philosophy for Our Time? *The Ecologist* 14(5/6), pp. 194–200.

Glacken, C. (1967) *Traces on the Rhodian Shore*. University of California Press, Berkeley.

Gullvag, I. & Wetleson, J. (eds.) (1982) *In Sceptical Wonder: Inquiries into the Philosophy of Arne Naess on the Occasion of His 70th Birthday*. Universitetsforlaget, Oslo.

Lovejoy, A. (1936). *The Great Chain of Being*. Harvard University Press, Cambridge, MA.

Naess, A. (1973) The Shallow and the Deep, Long-Range Ecology Movements: A Summary. *Inquiry* 16, pp. 95–100.

Naess, A. (1974) *Gandhi and Group Conflict: An Exploration of Satyagraha, Theoretical Background*. Universitetsforlaget, Oslo.

Naess, A. (1975) *Freedom, Emotion, and Self-Subsistence: A Systematization of a Part of Spinoza's Ethics*. Universitetsforlaget, Oslo.

Naess, A. (1979) Self-realization in Mixed Communities, of Humans,

Bears, Sheep, and Wolves. *Inquiry* 22, pp. 231–241.

Naess, A. (1984) A Defense of the Deep Ecology Movement. *Environmental Ethics* 6(3), pp. 265–70.

Naess, A. (1985) Identification as a Source for Deep Ecological Attitudes. In: Tobias, M. (ed.) *Deep Ecology*. Avant Books, San Diego.

Naess, A. (1986) Intrinsic Value—Will the Defenders of Nature Please Rise? In: Soulé, M. (ed.) *Conservation Biology: Science and Scarcity and Diversity*. Sinauer Associates, Sunderland, MA.

Pepper, D. (1984) *The Roots of Modern Environmentalism*. Croom Helm, London.

Reed, P. & Rothenberg, D. (eds.) (1987) *Wisdom and the Open Air: Selections from Norwegian Ecophilosophy*. Council for Environmental Studies, Univ. of Oslo, Norway.

Seamon, D. & Mugerauer, R. (eds.) (1986) *Dwelling, Place, and Environment*. Martinus Nighoff, The Hague.

Sessions, G. (1981) Shallow and Deep Ecology: A Review of the Philosophical Literature. In: Schulz and Hughes (eds.) *Ecological Consciousness*. University Press of America, Washington.

Sylvan, R. (1985) A Critique of Deep Ecology, Discussion Paper in Environmental Philosophy No. 12, Philosophy Department, the Australian National University, Canberra.

Watson, R. (1983) A Critique of Anti-Anthropocentric Biocentrism. *Environmental Ethics*, Vol. 5, pp. 245–256.

Watson, R. (1985) Challenging the Underlying Dogmas of Environmentalism. *Whole Earth Review* 45, pp. 5–13.

Section III

Major Topics

13

Feminism, Deep Ecology, and Environmental Ethics

Michael E. Zimmerman

Introduction

In this essay, I examine the feminist contention that both reform environmentalism and deep ecology are inadequate means for ending the human domination of nature, because both approaches ignore the decisive phenomena of patriarchalism and androcentrism. For example, one of the most important of the reform movements in environmental ethics urges us to "extend" rights to nonhuman beings in order to protect them from human abuse.[1] For many feminists, however, the concept of rights is so bound up with a masculinist interpretation of self and reality that it cannot serve to end the exploitation of nature that arises from that interpretation. At first glance, deep ecology would appear to be in agreement with the feminist critique of reformist environmental ethics. Deep ecology maintains that the humanity-nature relation cannot be transformed by moral "extensionism" or any other variety of reformism.[2] Instead, this transformation can only begin with the elimination of the anthropocentric world view that portrays humanity itself as the source of all value and that depicts nature solely as raw material for human purposes. Feminists claim, however, that deep ecology obscures the crucial issue by talking about *human*-centeredness, instead of about male-centeredness (*androcentrism*).[3] A truly "deep" ecology would have to be informed by the insights of eco-feminists, who link the male domination of nature with the male domination of woman. As Ariel Kay Salleh remarks, in deep ecology:

There is a concerted effort to rethink Western metaphysics, epistemology, and ethics..., but this "rethink" remains an idealism closed in on itself because it fails to face up to the uncomfortable psychosexual origins of our culture and its crisis.... Sadly, from the eco-feminist point of view, deep ecology is simply another self-congratulatory reformist move: the transvaluation of values it claims for itself is quite peripheral.... [T]he deep ecology movement will not truly happen until men are brave enough to rediscover and to love the woman inside themselves.[4]

In what follows I first present a brief account of the feminist explanation for the male domination of nature and woman. This account provides a context helpful for seeing why feminists are dissatisfied with the programs that some males have offered for healing the humanity-nature relationship. I then discuss, in turn, feminist critiques of the reformist "moral extensionism" and of deep ecology. Finally, I offer some critical reflections about some aspects of the feminist critique of deep ecology.

The Feminist Critique of the Domination of Nature

Contemporary feminism is a complex movement, with its share of disagreement about the origins, nature, and importance of patriarchy and androcentrism.[5] Most feminists would agree, however, that a major source of contemporary social and environmental ills is the fact that patriarchal culture has, on the one hand, repressed and devalued female experience and, on the other hand, has both absolutized and universalized male experience.... Ending this male domination could have dramatic consequences. According to Virginia Held, "If feminists can succeed not only in making visible but also in keeping within our awareness the aspects of 'mankind' that have been obscured and misrepresented by taking the 'human' to be the masculine, virtually all existing thought may be turned on its head."[6] So long as patriarchally raised men fear and hate women, and so long as men conceive of nature as female, men will continue in their attempts to deny what they consider to be the feminine/natural within themselves and to control what they regard as the feminine/natural outside themselves.

Marilyn French offers the following account of the origin of the male aversion to woman and nature.[7] Thousands of years ago, men gradually began to define the male as truly human, in contrast with the female, who was portrayed as being only partly human insofar as she was so closely identified with natural processes (birth, lactation, child rearing, menstrual cycles, etc.). The discovery by men of their role in pregnancy may also have enabled them to conceive of women no longer as a miraculous source of life from within themselves, but instead as mere carriers and nourishers of the seed implanted in them by men. The experience of being able to create something with so little personal involvement may have, in turn, led men to conceive of God as a transcendent, nonnatural, "male" source of power. This God replaced the Goddess who had emphasized pleasure, affiliation, mutual caring, harmony between nature and humanity. The idea of God emphasized power, hierarchy, independence, and dualism between nature and males. According to French, the rise of the transcendent, male power-god symbolizes the beginning of the human worship of power. She summarizes her views in the following way:

> Patriarchy is an ideology founded on the assumption that man is distinct from the animal and superior to it. The basis for this superiority is man's contact with a higher power/knowledge called god, reason, or control. The reason for man's existence is to shed all animal residue and realize fully his "divine" nature, the part that *seems* unlike any part owned by animals—mind, spirit, or control. In the process of achieving this, man has attempted to subdue nature both outside and inside himself; he has created a substitute environment in which he appears to be no longer dependent upon nature. The aim of the most influential human minds has been to create an entirely factitious world, a world dominated by man, the one creature in control of his own destiny. This world, if complete, would be *entirely* in man's control..., and man himself would have eradicated or concealed his basic bodily and emotional bonds to nature.[8]

The male's conception of himself as essentially cultural, nonfemale, nonnatural, immortal, and transcendent, as opposed to the essentially natural, noncultural, mortal woman, has continued in various guises for several thousand years.[9] Carolyn Merchant maintains, however, that the patriarchal view of nature as fearsome, threatening, wild, and uncontrollable was tempered for a long time by the alternative vision of Mother Nature: bounteous, kind, life-giving. But with the coming of the modern age, the motherly dimension of nature was eclipsed by the fearsome vision of a wild woman who must be known (Bacon) in order to be controlled. It is probably no accident that the great witchcraft trials raged during the time Europe was making the transition to a mechanistic world view. About this transition, Merchant remarks that:

> The metaphor of the earth as a nurturing mother was gradually to vanish as a dominant image as the Scientific Revolution proceeded to mechanize and rationalize the world view. The second image, nature as disorder, called forth an important modern idea, that of power over nature. Two new ideas, those of mechanism and of the domination and mastery of nature, became core concepts of the modern world.... As Western culture became increasingly mechanized in the 1600s, the female earth and virgin earth were subdued by the machine.[10]

At the basis of man's attempt to control nature is what has been called the "the God project": the quest to become divine, immortal, incorruptible.[11] The drive for such immortality may be said to motivate both the technological domination of nature as well as the nuclear arms race, either of which may result in the destruction of the Earth.[12] Men at war project death and evil onto the enemy: both sides are under the illusion that by killing the enemy, they will have eradicated both mortality and wickedness. Paradoxically, male humanity's effort to deny death and to control all things seems to hasten the death of all things. Rosemary Radford Reuther writes that:

> It is not extreme to see this [self-destructive] denouement as inherent in the fundamental patriarchal revolution of consciousness that sought to deny that the spiritual component of humanity was a dimension of the maternal matrix of being.... Patriarchal religion split apart from the dialectical unities of mother religion into absolute dualism, elevating a male-identified consciousness to

transcendent apriority. Fundamentally this is rooted in an effort to deny one's own mortality, to identify essential (male) humanity with a transcendent divine sphere beyond the matrix of coming-to-be-and-passing-away.[13]

While Judaeo-Christian scripture sometimes accords nature goodness insofar as it is a creature of God, more often these scriptures assert the absolute difference between God and creation.[14] Genesis tells us that only humans are made "in the image of God" and, hence, are given dominion over the rest of creation. In the patriarchal culture of Jews and Christians, this idea of human dominion over creation was conceived as *male* dominion. In early modern times, the view of the special status of humanity in general, and males in particular, was secularized. Today, even nonreligious modern people take it for granted that there is a natural hierarchy at the top of which stands humankind.[15] Modern Western humanity presumes that only humans are the source of truth, value, and meaning; nature is merely an object whose sole value lies in its usefulness for man. Nature must be channeled and repressed for the purpose of human control, security, and survival. In industrial society, men are trained and disciplined in ways that repress the "useless" and "counterproductive" aspects of nature at work in them, including feelings, emotions, and other "womanly" sensibilities. Power over the human organism is a crucial ingredient of the technological domination of the rest of nature.[16]

The technological project is closely linked to the scientific revolution initiated by thinkers such as Descartes. Descartes' extreme rationalism and his subject-object dualism are the products of an extremely masculinist view of self and reality, a view that is shared by many males in modern society. Cut off from their feelings, men become isolated, rigid, overly rational, and committed to abstract principles at the expense of concrete personal relationships. As a result of their attachment to abstract doctrines, males have developed highly rationalistic moral philosophies. Such philosophies include little or no role for *caring* and *feeling* as preconditions for ethics, including the ethic concerning humanity's relation to nature. Marti Kheel notes:

What seems to be lacking in much of the literature in environmental ethics (and in ethics in general) is the open admission that we cannot even begin to talk about the issue of ethics unless we admit that we *care* (or *feel something*). And it is here that the emphasis of many feminists on personal experience and emotion has much to offer in the way of reformulating our traditional notion of ethics.[17]

Feminists maintain that most modern moral theory is linked to the very androcentric-patriarchal way of thinking that is responsible for the domination of nature. The rationalistic subject-object dualism is mirrored in the abstract, calculative, rational, and atomistic ethical systems that have arisen to govern competition among men after the death of their biblical God. Such systems lack the relational-intuitive sensibility that feminists maintain is required for the new *ethos* in which the nature-humanity dualism is overcome. The doctrine of "the natural rights of man" is allegedly an example of such an androcentric ethical system. In recent years, a number of environmental philosophers have attempted to "extend" rights to nonhuman beings, in order to protect those beings from human abuse. In the next section, I examine the feminist contention that such moral reformism cannot heal the nature-humanity dualism, since the concept of "rights" is linked to the rationalistic-atomistic metaphysics that has led to the domination of nature.

The Feminist Critique of the "Natural Rights of Man"

It has been argued that moral philosophers attempting to "expand" prevailing moral categories to "cover" the cases of animals or plants are analogous to the "normal scientists" trying to shore up the prevailing paradigm in the face of anomalies (cf. Kuhn).[18] In the face of the anomalous moral issues involved in the human exploitation of nature, however, refinements of the existing ethical paradigms will not be of much help. What is needed, we are told, is a "paradigm shift" that produces a nonandrocentric framework in which nature will appear as something other than an object to be dominated. Such a shift is *not* involved in the efforts by some reformers to "extend" rights to nonhuman beings in order to protect them from human abuse.

Natural rights theory is a typical example of a masculinist moral system, i.e., a system based on a male way of perceiving self, other, and nature.[19] The fundamental claims of natural rights theory are: that humans are endowed with certain inalienable rights, including life, liberty, and property; that humans are morally prohibited from interfering with another person's rights so long as that person does not interfere with legitimate rights of their own; and that humans have a right to defend their rights against those who would attempt wrongfully to deprive them of their rights. Natural rights theorists tend to restrict rights (and moral standing) to human beings, and they portray nonhuman beings as being virtually devoid of intrinsic worth. Locke argued that labor ontologically transforms natural things; they go from being valueless entities to valuable ones when human labor is mixed with them.[20] The fact that Locke is one of the founders of liberal capitalism, and that his labor theory of value was very influential on Karl Marx, helps to explain why both capitalism and communism treat nature primarily as an object for human use.[21]

Feminists are not alone in criticizing the doctrine of natural rights, but they maintain that only the feminist critique enables us to grasp both the origins and impact of the atomistic and egoist self-understanding lying at the base of the "rights of man." The doctrine of natural rights, we are told, is androcentric, hierarchical, dualistic, atomistic, and abstract. It is *androcentric* because its conception of human beings is based on a masculinist experience which excludes (and implicitly negates) female experience; it is *hierarchical* because it gives preference to male experience, and also because it portrays humans as radically more important than any other sort of beings; it is *dualistic* because of its distinction between humans (rational, intrinsically valuable, rights-possessing) and nonhumans (nonrational, instrumentally valuable, rights-lacking); it is *atomistic* because its portrayal of human beings (as separate egos) is consistent with the atomistic metaphysics of modern science; and it is *abstract* because conflicts about rights are resolved in rationalistic and impersonal terms that ignore both the feelings and the particular traits/needs of the individuals involved. Founders of the doctrine of natural rights were men who presupposed that the male experience of self, world, and morality was universal. According to Naomi Scheman, this view is perpetuated by the way in which males are raised within patriarchal structure:

> The view of a separate, autonomous, sharply individuated self embedded in liberal political and economic ideology and in the individualist philosophies of mind can be seen as a defensive reification of the process of ego development in males raised by women in a patriarchal society. Patriarchal family structure tends to produce men of whom these political and philosophical views seem factually descriptive and who are, moreover, deeply motivated to accept the truth of those views as the truth about themselves.[22]

The idea of the self as an isolated ego competing with other egos for scarce resources was most forcibly articulated by Hobbes. Later, this idea was reinforced by Darwin's evolutionary doctrine, which itself seems to reflect his own experience of competitive, egoistical social relations in nineteenth-century England.[23] Thus, the theory that competitive behavior is "natural" and "necessary" seems based at least in part on the fact that Darwin saw nature through the framework of competitive social categories. It has been argued that the "state of nature" was neither so bellicose nor so male-dominated as Hobbes and Darwin would have had us believe.[24] So long as people conceive of themselves as isolated, autonomous egos, who are only externally related to others and to nature, they inevitably tend to see life in terms of scarcity and competition. When people conceive of themselves as internally related to others and to nature, however, they tend to see life in terms of bounty, not scarcity; cooperation, not aggressive competition.

The isolationist-competitive view of human nature, then, reflects not a fact about human nature, but instead the experience of men raised in a patriarchal culture. Feminists argue that liberal political philosophers can adhere to a doctrine of the isolated, independent male ego because they presuppose that *women at home* will continue to knit together the social fabric on which the competing male egos depend.[25] Scheman maintains that:

Men have been free to imagine themselves as self-defining only because women have held the intimate social world together, in part by seeing ourselves as inseparable from it. The norms of personhood, which liberals would strive to make as genuinely universal as they now only pretend to be, depend in fact on their not being so—just what we should expect from an ideology.[26]

From the masculinist perspective, the self appears not to be constituted by relationships with others, but instead is a self-contained entity which constitutes temporary, external relationships with other self-contained entities. If a relationship is terminated (that is, if a contract is ended or broken), the self-contained self is not changed, since the relationship is wholly external. According to Nancy Chodorow, a leading proponent of the psychological school called "object relations" theory, this male view of the self as an isolated ego stems from early childhood relations between son and mother.[27] At first, the little boy identifies himself with his mother; later, however, he discovers that he is sexually differentiated from her. Seeking to gain his own sexual identity, the boy experiences his own withdrawal from his mother as abandonment. Experiencing profound anger and grief because of this perceived abandonment, he subsequently fears, mistrusts, and hates women. Moreover, he tends to define himself in a negative way as *not female*. His fear and anger lead him to want to dominate both the woman (mother image) within himself and the woman outside of him. As he grows up, he shields himself from his feelings, which overwhelmed him during the trauma of separation, and he defines himself as radically separate from others: he is hesitant to involve himself once again in relationship, since his most important relationship ended in such pain. Because "mother" was originally identified with all of reality, boys and men tend to regard as "female" the undifferentiated natural background against which individual entities stand out. Mother Nature, then, appears as a threatening, unpredictable force from which a man must differentiate himself and which he must control. Because girls maintain their sense of identity with their mothers for a much longer time than do boys, their sense of self is bound up with relationship. Many women claim that they do not experience themselves as radically separate, self-contained egos, but instead as a network of personal relationships. These relationships are not external, but internal and constitutive of the "self." If a relationship is removed or disrupted, the "self" is inevitably affected. Hence, the notion of fending only for "oneself" does not have the same persuasiveness for many women as it tends to have for men, who conceive of themselves as essentially unrelated to others. Caring for others is, for a woman, difficult to distinguish from caring for herself. If men often have difficulty in relating to others, women often have difficulty in assuming their own identity. Carol Gilligan has postulated that the differing senses of self possessed by males and females lead to differences both in moral perception and moral decision making.[28] Hence, the male preoccupation with "rights" can be related to the male sense of being an isolated ego competing with other egos for ostensibly scarce resources.

When social atomism was being developed in the seventeenth century, it reflected the scientific trend to conceive of material reality atomistically. Hobbes, for example, explicitly modeled his philosophical anthropology on mechanistic-atomistic scientific principles. Most other thinkers, however, were less strong-minded than Hobbes; partly because they clung to religious ideas about man's immortal soul, they hesitated to reduce man to the state of a mere machine. Thus, they employed Cartesian dualism to distinguish "rational" humanity from "extended" material reality. This dualism allowed them to depict nature as a mere thing without intrinsic value, since nature lacks mind or soul. The doctrine that human beings are intrinsically valuable because they alone possess an immortal soul is based on ingredients drawn from the Greek and Judaeo-Christian traditions. Even after the decline of Greek metaphysics and the Judaeo-Christian tradition, modern people continue to maintain that human beings are somehow "special."

This continuing sense of human specialness played an important role in the development of modern moral and legal philosophy, which is notoriously anthropocentric. Hugo Grotius, for example, transformed the Roman doctrine of *jus naturae* (natural right or law) so that it no longer applied to all creatures, but only to rational, self-interested creatures capable of entering into contracts.[29] Plants and animals, lacking such capacity, were said not to have any rights. Kant, moreover, argued that animals lack moral standing because they are not rational; we are prohibited from abusing them only because such practices may encourage humans to abuse each other.

In part, these changes in moral and legal philosophy stemmed from the modern scientific view that the universe is without meaning or purpose, that it is composed of externally related atoms, and that the atomistic human ego is the source of purpose, value, and meaning. Today, however, many scientists no longer view reality as being constituted by individual, isolated, externally related entities, but instead as a network of internal relationships.[30] It is, then, an example of what Whitehead called "misplaced concreteness" to define an animal apart from the environment (air, land, trees, food chains, predators, etc.) that constitutes its "niche." Nothing is separate; all is connected internally. Internal relations are such that they constitute the "reality" of the thing in question; if you alter the relationships, you alter the thing, since it does not exist apart from those relationships. Humanity, too, is an aspect of the fabric of life on Earth; we are not apart. Hence, the emergence of self-conscious human beings can be interpreted as an event by which nature can observe and evaluate *itself.*

The fact-value distinction so important for modern science and philosophy may be undermined by the realization that the universe itself generates novelty, purpose, and diversity. The current dialogue between science and theology reflects the growing trend to regard categories such as "meaning" and "purpose" not as anthropomorphic projections, but instead as ingredients in the cosmos.[31] Furthermore, the subject-object dualism lying at the base of the fact-value dichotomy is now being overcome by significant changes in the idea of what constitutes scientific research. The masculinist conception of science as an abstract and rationalistic quest for the universal ignores the fact that the quest for understanding must involve moods and feelings that disclose crucial aspects of the particular and unique.[32] Androcentric thinking denigrates feelings by asserting that they lack the persuasiveness and universality of conclusions arrived at by way of rational argumentation. Supposedly, rational calculation alone can serve as the proper guide for the ego in its quest for survival in the harsh, competitive world. From this viewpoint, the only limits to humanity's use of nature are *prudential* ones. Before our enormous industrial activity causes the biosphere to collapse, market mechanisms and technological innovations will lead self-interested human beings to adjust their behavior in order to preserve the conditions needed for human life. Yet the purely prudential and self-interested calculations at work in the exploitation of the nonhuman world can go forward smoothly only so long as one does not allow oneself to *feel* the consequences those calculations have for life on Earth—human as well as nonhuman.[33]

In summary, while many feminists acknowledge that the concept of human rights reveals a genuine concern for respecting the interests of other individuals, many feminists are critical of the concept because it is based on masculinist experience that is wrongly universalized and because it fails to include moral categories that arise from a feminine experience of self and world. The experience of relatedness reported by many women gives rise to a morality of caring for the concrete needs of those with whom one is related. This sense of concrete relationship and kinship extends to the natural world as well. Hence, feminists argue that the domination of nature is a masculinist project, one rooted in man's disassociation from the natural world. The doctrine of natural rights is unsuitable for establishing a nondomineering relation between humanity and nature because it (1) is androcentric, (2) regards nonhuman beings as having only instrumental value, (3) is hierarchical, (4) is dualistic, (5) is atomistic, (6) adheres to abstract ethical principles that overemphasize the importance of the isolated individual, (7) denies the importance of feeling for informing moral behavior, and (8) fails to see the essential relatedness of human life with the biosphere that gave us birth. Hence, environmental ethicists who hope to protect nature by "extending" rights to nonhuman beings are part of a reform movement that cannot succeed.

Before concluding this section, some critical observations are in order. After having gone through the phase of seeking to dissolve differences between men and women, many feminists began to affirm those differences—and to conclude that woman is better than man.[34] Hence, some feminists have praised relatedness and feelings at the expense of allegedly "masculinist" traits of individuality and rationality. But other feminists warn that this move runs the risk of simply reaffirming traditional views that women are "feelers," while men are "thinkers." Affirmation of such views is rooted in "essentialist" doctrines of the differences between men and women. Feminists opposed to such essentialism argue that reasoning and feeling are *human* capabilities that do not belong exclusively to one sex or the other. In a non-patriarchal society, human beings would presumably manifest a healthy interplay between emotion and thinking—and moral issues would be informed by both as well. Yet, the notion that a healthy human being would be androgynous, that is, a "combination" of traits currently described as "male" or "female," is problematic insofar as that notion maintains the dualism between male and female. At this stage in human history, we are still groping to understand what it would mean to be a mature man or woman in a non-patriarchal society.[35]

In any event, many feminists are cautious about simply rejecting the morality of rights and replacing it with a morality of feelings. According to Carole Pateman, for example, some feminists have argued that "since 'justice' is the work of men and an aspect of the domination of women, women should reject it totally and remake their lives on the basis of love, sentiment, and personal relations."[36] Pateman counters by arguing that the liberal concepts of rights, justice, and the individual help guide the dialectic that goes on "between the particular or personal and the universal or political. . . ."[37]

Carol Gilligan has also suggested that a morality of compassion based on the feminine sense of relatedness is complementary with the morality of justice based on the masculine sense of separateness. By overemphasizing interrelatedness, feminists run the risk of leaving no categories for conceiving of people as individuals, or for making moral choices when faced with conflicts between individuals. As Jane Flax notes, "Women, in part because of their own history as daughters, have problems with differentiation and the development of a true self and reciprocal relations."[38]

Overemphasizing internal relatedness can also be a problem when it comes to environmental ethics. Marti Kheel warns of the danger of a kind of environmental totalitarianism that sacrifices the individual for the good of the whole. Kheel insists that while individuals are not radically separate, but instead are internally related "knots" within the fabric of reality, these knots are intrinsically important:

> A vision of nature that perceives value both in the individual and in the whole of which it is a part is a vision that entails a reclaiming of the term *holism* from those for whom it signifies a new form of hierarchy (namely, a valuing of the whole over the individual). Such a vision asks us to abandon the dualistic way of thinking that sees value as inherently exclusive (i.e., they believe that the value of the whole cannot also be the value of the individual).[39]

Thinkers such as Pateman and Kheel help justify the conclusion that the doctrine of natural rights may be useful if applied in a nonpatriarchal, nonatomistic view of humanity and nature, i.e., a view which emphasizes both the essential interrelatedness of all things and the concrete character of the relations between individual "knots" in the cosmic whole. However, so long as natural rights theory presupposes that possessors of rights must be self-interested, rational agents capable of fulfilling the "duties" corresponding to rights; and moreover, so long as natural rights theory clings to metaphysical atomism and egoism, then natural rights theory will not be very useful in correcting current moral problems in the humanity-nature relationship. The reformist impulse behind the extension of rights to nonhuman beings must transform itself into a profound critique of the metaphysical and epistemological presuppositions of patriarchal culture. These androcentric and anthropocentric presuppositions blind us to the fact that human beings are not radically separate from nature, but instead are manifestations of it. We tend to assign rights to those beings which possess attributes (such as consciousness or awareness) resembling our own, for we assume that our attributes are the measure for the rest of reality. A more humble humanity attuned to the internal relatedness of all things would presumably respect all things as ingredients in a social cosmos.

Radical Feminism and Deep Ecology

At first glance, deep ecology may seem to be in almost complete agreement with the feminist view that abstract, dualistic, atomistic, and hierarchical categories are responsible for the domination of nature. A new *ethos*, according to deep ecologists, is required for humans to dwell appropriately on Earth. Moreover deep ecologists—like feminists—have been critical of reformist attempts to extend modern moral categories to "protect" nonhuman beings from human abuse. Feminist critics of deep ecology, however, assert that it speaks of a gender-neutral "anthropocentrism" as the root of the domination of nature, when in fact androcentrism is the real root. Only the interpretive lens of androcentrism enables us to understand the origin and scope of dualistic, atomistic, hierarchical, and mechanistic categories. Deep ecologists are still only reformists: they want to improve the humanity-nature relationship without taking the radical step of eliminating both man's domination of woman (including the woman inside of each man) and the culturally enforced self-denigration of woman. Moreover, since deep ecology was formulated almost exclusively by men, and since men under patriarchy allegedly think in distorted ways, the similarity between the principles of deep ecology and what we might call eco-feminism may be largely superficial.

In her article "Deeper than Deep Ecology: The Eco-Feminist Connection," Ariel Salleh makes a number of specific criticisms of deep ecology, especially certain works by Arne Naess and Bill Devall.[40] First, deep ecologists often use sexist language, as for example when they speak unself-consciously of the need for improving "man's relation with nature." Use of such language reveals that deep ecologists have not acknowledged the basic social inequality between men and women. Salleh argues that "The master-slave role which marks man's relation with nature is replicated in man's relation with woman." Deep ecologists argue that artificial control of population is a necessary means for developing a new relation with nature. Here, deep ecologists seem to preach the same gospel as other men before them: controlling female reproductive processes by technical means will solve problems allegedly caused by a natural process. Deep ecologists use highly rationalistic arguments that betray their ongoing commitment to masculinist-scientistic modes of thought. Because their experience is deformed by masculinist modes of thought, male deep ecologists should consult women who are more in tune with the natural world than are men, and who are open to the experience of reality in an alternative way. Solutions proposed by deep ecologists are naive, according to Salleh, insofar as they are offered outside of the context of the critique of patriarchalism. For example, when deep ecologists call for decentralizing society, they ignore the fact that patriarchal culture has always favored hierarchy and centralization—and that unless patriarchal consciousness is abandoned, schemes for decentralization are hopeless. Despite their good intentions, then, deep ecologists exhibit a pervasive masculinist bias that works against their aims. Salleh says that

> In arguing for an eco-phenomenology, [Bill] Devall certainly attempts to bypass this ideological noose [mechanistic metaphysics]—"Let us think like a mountain," he says—but again the analysis here rests on what is called a gestalt of person-in-nature: a conceptual effort, a grim intellectual determination to care; to show reverence for Earth's household and "to let" nature follow "its separate" evolutionary path.[41]

Salleh concludes that deep ecologists are males who, damaged by patriarchy, are seeking to heal themselves:

> Watts, Snyder, Devall, all want education for the spiritual development of "personhood." This is the self-estranged male reaching for the original androgynous natural unity within himself. The deep ecology movement is very much a spiritual search for people in a barren secular age; but how much of this quest for self-realization is driven by ego and will? If, on the one hand, the search seems to be stuck at an abstract cognitive level, on the other, it may be led full circle and sabotaged by the ancient compulsion to fabricate perfectibility. Men's ungrounded restless search

for the alienated Other part of themselves has led to a society where not life itself, but "change," bigger and better, whiter than white, has become the consumptive end. . . . But the deep ecology movement will not truly happen until men are brave enough to rediscover and to love the woman inside themselves. And we women, too, have to be allowed to love what we are, if we are to make a better world.[42]

Salleh's critique is, in my opinion, only partly accurate, and her accusatory tone may limit her audience as much as the misogyny of a great deal of systematic thinking diminishes its applicability. It may well be, of course, that men—especially those men who are seeking to move beyond the constricted categories of modern manhood—need to experience the righteous anger of women who have for so long experienced the repressiveness of patriarchal society. Salleh is right, moreover, in saying that most deep ecologists continue to write in the technical-rationalistic style that gives their work some measure of credibility within patriarchy. Yet feminists themselves are familiar with the problem of discovering their own "voice." And Salleh herself uses a style of writing and argumentation that does not seem radically different from that of deep ecologists such as Devall or Naess. The fact is that in order to gain a hearing within "establishment" journals and presses, authors (male and female alike) must conform to traditional linguistic forms, even though those forms may be aligned with patriarchal social structures.

It is not surprising, of course, to hear that deep ecologists tend to write in ways that are called masculinist, since this is how men and women alike are socialized to write. Is it not possible, however, that despite such a writing style, and despite how they've been socialized, male deep ecologists may in fact open to their own feelings and to their relatedness to nature in ways that evade the effects of patriarchy? What to women might appear as a clumsy, obviously male way of speaking might be for the male speaker an expression of a genuine sense of kinship with nature, including the nature within him.

Is it not too sweeping a generalization to say that women are more attuned to nature than are men? Not for Salleh, who claims that deep ecologists are attuned to nature only in a manner distorted by patriarchal culture, and that their masculinist forms of speaking and writing are signs of that distorted experience. She asserts that such distorted experience occurs because deep ecologists are out of touch with the woman within them. Is the term *woman* meant to refer to the feelings, emotions, and relational sensibility with which many men are out of touch? Yet to conceive of such traits as "feminine" seems to suggest an essentialist and/or genetic doctrine of the differences between men and women: that man is thinker, woman is feeler. Is such a doctrine consistent with the conviction of many feminists that men and women alike are distorted products of the psychological, social, and cultural practices of patriarchy? If we humans are essentially or naturally dichotomized by sex-linked traits (reason vs. feeling), then there is no real point in trying to change human cultural practices. In recent years, a number of feminists have favored such an essentialist view and have concluded that woman is better than man.

This variety of essentialist feminism often rejects reason, science, technology, abstraction, individuation, hierarchy, and so on, as the bitter fruits of patriarchal culture. Other feminists, however, argue that such categories and practices are not intrinsically evil; instead, they become destructive when utilized in a one-sided culture such as patriarchy. The fact that these categories arose within the history of patriarchy does *not* mean that truly liberated women should eschew them entirely in order to avoid being "infected" with patriarchalism. Great scientists report that their work is not merely rational and deductive, but involves insight into particular relationships and concrete events. Modern science has gone astray by its pretense to pure objectivity and rationality. Contemporary feminist criticism may help bring about a change in our understanding of the nature of scientific inquiry. Modern science and technology are potentially liberating, but have been misdirected because of the patriarchal view that nature (including human nature) must be exploited to enhance power and security. When social categories change, feminists argue, the technological domination of nature will change accordingly.

The fact that all people tend to be distorted under patriarchal culture leads to another observation about Salleh's critique of deep ecology. If deep ecologists cannot get to the heart of the matter because their experience is too deeply distorted by patriarchy, cannot we say something analogous about women? How can authentic female experience and self-expression be possible under patriarchy? And what can be meant by the concept of "authentic" female experience? Leaving aside essentialist arguments, we would have to conclude that authentic experience cannot be identical with "natural" experience, since human experience is always culturally mediated. Would authentic female experience be that formed by a feminist culture? But what then of authentic male experience? Would it be possible in such a culture? Or must there be two separate cultures, male and female, for members of the two sexes to become truly human? If patriarchy is an interpretive framework, is feminism itself not another such framework? Does feminism pretend to provide a nondistorted, impartial way of interpreting experience? Are feminists raised under patriarchy motivated by their own version of the power drive that is essential to patriarchy?

To such questions, feminists might reply in a two-fold way. First, the quest to articulate feminine experience is motivated by the conviction that many women experience self and world in a way differently than is reported by men. Women want and need to validate the fact that there is *another mode* of experience. Such validation is important, even if the experience of women is distorted by patriarchy. Second, this validation makes possible the next stage of development: the search not simply for authentic female experience, but for authentic *human* experience. Required for such experience are cultural arrangements and categories compatible with the development of people who are aware both of their own uniqueness, and of their profound relationships with other people and the rest of the natural world. Cultural practices, then, not genetic differences, are responsible for the current differentiation between "masculine" and "feminine" experience. Ultimately, then, feminism seeks to liberate women and men alike from the distorted mode of existence necessitated under patriarchy. Authentic human existence would inevitably transform the current exploitative treatment of nature.

Critics of feminism, however, regard as disingenuous the claim that the real motive of feminism is to liberate *all* people. Such critics contend that feminists have their own power agenda. Feminists make patriarchy responsible for too much; they portray men as the villains of world history, even though some feminists try to temper this portrayal by saying that *individual* men are not to blame, since they have been socialized to behave in a domineering and destructive way. Further, radical feminism accuses men of projecting unwanted traits onto women and nature. Yet projection is surely not something in which men alone engage. What traits, then, are women projecting onto men? And what benefits accrue to women through projecting such traits? Do women split off from themselves and project onto men violence, aggressiveness, selfishness, greed, anger, hostility, death hating, nature fearing, individuality, and responsibility? And as a result of bearing these projected traits, do men behave much more violently, selfishly, etc., than they would if these projected traits were withdrawn by women? If women do project these traits, one benefit gained would be for them to regard themselves as peaceful, charitable, concerned about others, compassionate, emotional, in harmony with nature, loving, thoughtful, and more truly human. But can such positive, "good" characteristics belong only to one sex? In searching for their own "voice," are feminists willing to acknowledge their own "dark side," which is all too easily projected onto men? And do they realize that men, too, are the victims of patriarchy, that they lack a real "voice" of their own, apart from the impersonal voice that they have assumed in the process of having to split off their own feelings? While benefiting from the material well-being and technological progress made possible by masculinist science and industry, do women rid themselves of responsibility for the negative side effects of such progress by attributing them to rapacious male behavior? Feminists can say that patriarchal categories are the problem, and that changing those categories according to feminist principles will bring about an end to the domination of woman and nature. Yet there is no assurance that new forms of domination and power will not arise in the process.

By way of reply, many feminists acknowledge that individual men turn out as they do as a result of ancient child-rearing and social-ization processes that they themselves did not choose, that women must own up to what they project onto men, that women have their own fears about mortality and have their own resentment against nature, and that the feminist critique of patriarchy must engage in searching self-criticism. Feminists would remind critics, however, of the danger involved in *blaming the victim* for her present condition. For example, in medieval China mothers were responsible for enforc-ing upon their daughters the terrible practice of foot binding. It is wrong to conclude, however, that women were responsible for this practice, since they were part of a patriarchal culture which *expected* them to behave in this way. Moreover, the fact that women tend to split off from themselves and to project "masculine" traits onto men tends to *disempower* women, and is encouraged by men. Women suffer from being the "second sex." Finally, feminists insist that they are not interested in instituting new forms of control and domina-tion, but instead are seeking to design a participatory process that empowers women and men alike. At present, many people find it difficult to imagine an alternative to a society based on hierarchy and control, since this is the only sort of society we have known for centuries. Human imagination, courage, commitment, and lots of time will be necessary to bring to fruition the dream of a process-oriented society.

To the extent that some women have been less socialized by the atomistic, dualistic, hierarchical categories that—when employed under patriarchy—appear to be responsible for much ill-treatment of nature and woman, it is plausible to suggest that those women are in a better position than most men to help reconstruct the human-ity-nature relation in light of their ongoing sensitivity toward and involvement with their own bodies and the rest of nature. We must be careful, however, not to fall prey to the sex-based stereotyping that has been so crucial to maintaining patriarchy. Men and women are both capable of becoming more open to and at harmony with the natural world. Deep ecologists and eco-feminists need to unite in reconstructing Western humanity's current attitudes toward nature.

Notes

1. Examples of such efforts include Christopher D. Stone, *Should Trees Have Standing? Toward Legal Rights for Natural Objects* (Los Altos, CA: William Kaufmann, 1974); Peter Singer, *Animal Liberation* (New York: New York Review, 1975); Joel Feinberg, "The Rights of Animals and Unborn Generations," in *Philosophy and Environmental Crisis*, ed. William T. Blackstone (Athens: University of Georgia Press, 1974).

2. Some important works on deep ecology include: Arne Naess, "The Shallow and the Deep, Long-Range Ecology Movement," *Inquiry* 16 (1973): 95–100; Arne Naess, "The Arrogance of Anti humanism?" *Ecophilosophy* 6 (1984); Naess, "Identification as a Source of Deep Ecological Attitudes," in *Deep Ecology*, ed. Michael Tobias (San Diego: Avant Books, 1984); George Sessions, "Shallow and Deep Ecology: A Review of the Philosophical Literature," in *Ecological Consciousness: Essays from the Earthday X Colloquium*, ed. Robert C. Schultz and J. Donald Hughes (Washington, DC: University Press of America, 1981); Sessions, "Ecological Consciousness and Paradigm Change," in *Deep Ecology*; William B. Devall, "The Deep Ecology Movement," *Natural Resources Journal* 20 (1980): 299–322; Bill Devall and George Sessions, *Deep Ecology: Living as if Nature Mattered* (Salt Lake City: Peregrine Smith Books, 1985); Warwick Fox, "Deep Ecology: Toward a New Philosophy for Our Time?" *The Ecologist* 14 (1984): 194–204; Alan Drengson, *Shifting Paradigms: From Technocrat to Planetary Person* (Victoria, BC: LightStar Press, 1983); Michael E. Zimmerman, "Toward a Heideggerean *Ethos* for Radical Environmentalism," *Environmental Ethics* 5 (1983): 99–131; Murray Bookchin, *The Ecology of Freedom: The Emergence and Dissolution of Hierarchy* (Palo Alto: Cheshire Books, 1982). Fritjof Capra and Charlene Spretnak, in *The Green Movement: The Global Promise* (New York: Dutton, 1984), provide a synthesis of deep ecology and radical feminism which can be called eco-feminism. In his book, *The Turning Point: Science, Society, and the Rising Culture* (New York: Bantam Books, 1982), Fritjof Capra argues that the decline of patriarchy is one of the three crucial changes that will usher in a new era of improved relations between humanity and the natural world.

3. For bibliographical help on the woman-nature relationship, cf. Jane Yen, "Women and Their Environment: A Bibliography for Research

and Teaching," *Environmental Review* 8 (1984): 86–94. Significant contributions to the literature include: Marti Kheel, "The Liberation of Nature: A Circular Affair," *Environmental Ethics* 7 (1985): 135–149; Ynestra King, "Toward an Ecological Feminism and a Feminist Ecology," in Joan Rothschild, ed., *Machina ex Dea: Feminist Perspectives on Technology* (New York: Pergamon Press, 1983); Carolyn Merchant, *The Death of Nature: Women, Ecology and the Scientific Revolution* (New York: Harper & Row, 1980); Annette Kolodny, *The Lay of the Land: Metaphor as History in American Life and Letters* (Chapel Hill: University of North Carolina Press, 1975); Kolodny, *The Land Before Her* (Chapel Hill: University of North Carolina Press, 1983); Brian Easlea, *Science and Sexual Oppression: Patriarchy's Confrontation with Woman and Nature* (London: Weidenfeld and Nicholson, 1981); Elizabeth Dodson Gray, *Green Paradise Lost* (Wellesley, MA: Roundtable Press, 1979).

4. Ariel Kay Salleh, "Deeper than Deep Ecology: The Eco-Feminist Connection," *Environmental Ethics* 6 (1984): 339–45.

5. In *Feminist Politics and Human Nature* (Totowa, NJ: Rowman & Allanheld, 1983), Alison M. Jaggar provides an insightful account of four major schools of feminism: liberal, traditional Marxist, radical, and socialist. Jaggar favors the Marxist approach. Radical feminists maintain that socialist, Marxist, and liberal feminists make use of masculinist political ideologies. A change of class structure, or the extension of rights to women, cannot solve the problems created by patriarchal consciousness. Other very helpful analyses of the history of modern feminism can be found in Hester Eisenstein, *Contemporary Feminist Thought* (Boston: G.K. Hall & Co., 1983); and Hester Eisenstein and Alice Jardine, eds., *The Future of Difference* (New Brunswick: Rutgers University Press, 1985).

6. Virginia Held, "Feminism and Epistemology: Recent Work on the Connection between Gender and Knowledge," *Philosophy and Public Affairs* 14 (1985): 296–307. For a thoughtful critical discussion of the radical feminist critique of masculinist epistemology, cf. Lorraine B. Code, "Is the Sex of the Knower Epistemologically Significant?" *Metaphilosophy* 12 (1981): 267–276.

7. Marilyn French, *Beyond Power: Women, Men, and Morality* (New York: Summit Books, 1985). According to Ellen Messer-Davidow (personal communication), "Feminist anthropologists point out that some peoples do not align male/female and culture/nature, a construct of the

Enlightenment..., but use generations, language proficiency, or residence (clearing and bush) as constructs to organize their cultures." The male/female, culture/language relationship, then, is apparently not as universal as French seems to think. For an insightful treatment, from a male perspective, of the love-hate relation men have with women, cf. Wolfgang Lederer, *The Fear of Women* (New York: Harcourt, Brace Jovanovich, 1968).

8. French, *Beyond Power*, p. 341.

9. On the patriarchal attitude toward nature, cf. Sherry B. Ortner, "Is Female to Male as Nature is to Culture?" in Michelle Rosaldo and Louise Lamphere, eds., *Woman, Culture, and Society* (Stanford: Stanford University Press, 1974); and Joan Bamberger, "The Myth of Matriarchy: Why Men Rule in Primitive Society," in *Woman, Culture, and Society*.

10. Carolyn Merchant, *The Death of Nature*, p. 2. Cf. also Brian Easlea, *Witch-Hunting, Magic, and the New Philosophy* (Atlantic Highlands, NJ: Humanities Press, 1980); and Mary Daly, *Gyn/Ecology* (Boston: Beacon Press, 1978).

11. Cf. R. Reuther, *New Woman, New Earth* (New York: Seabury Press, 1975); also cf. Ken Wilber, *The Atman Project* (Wheaton, IL: Quest Books, 1980); Wilber, *Up From Eden: A Transpersonal View of Evolution* (Boulder, CO: Shambhala, 1981).

12. Cf. Michael E. Zimmerman, "Humanism, Ontology, and the Nuclear Arms Race," *Research in Philosophy and Technology* 6 (1983): 151–72; Zimmerman, "Anthropocentric Humanism and the Arms Race," *Nuclear War: Philosophical Perspectives*, ed. Michael Fox and Leo Groarke (New York: Peter Lang Publishers, 1985).

13. Reuther, *New Woman, New Earth*, pp. 154 ff.

14. The pioneering essay criticizing the environmental consequences of the Judaeo-Christian tradition is Lynn White, Jr., "The Historical Roots of our Ecologic Crisis," *Science* 155 (1967): 1203–07. Cf. also David Crownfield, "The Curse of Abel," *North American Review*, Summer, 1973, pp. 58–63. Cf. Mary Daly's famous feminist critique of patriarchal Christianity, *Beyond God the Father* (Boston: Beacon Press, 1973); also Reuther, *New Woman, New Earth*. In defense of the Judaeo-Christian tradition, cf. Loren Wilkinson, *Earthkeeping: Christian Stewardship of Natural Resources* (Grand Rapids: Erdman's, 1980); Wesley Granberg-Michaelson, *A Worldly Spirituality: The Call to Redeem the*

Earth (New York: Harper & Row, 1985). Cf. also the work done by Vincent Rossi and his colleagues of The Eleventh Commandment Fellowship, PO Box 14606, San Francisco, CA 94114. The "eleventh commandment" reads: "The Earth is the Lord's and the Fullness thereof: Thou shalt not despoil the Earth, nor destroy the Life thereon."

15. Cf. William Leiss, *The Domination of Nature* (Boston: Beacon Press, 1972).

16. Concerning the idea that modern technology shapes both human body and mind to its purposes, cf. the work of members of the Frankfurt Critical School, such as Max Horkheimer, Theodor Adomo, and Herbert Marcuse; more recently, cf. the work of Michel Foucault; also cf. David Michael Levin, "The Body Politic: Political Economy and the Human Body," *Human Studies* 8 (1985): 235–78.

17. Kheel, "The Liberation of Nature," pp. 143–44.

18. Cf. Thomas Kuhn, *The Structure of Scientific Revolutions* (Chicago: University of Chicago Press, 1970).

19. For a review of criticisms of natural rights, cf. Michael E. Zimmerman, "The Crisis of Natural Rights and the Search for a Non-Anthropocentric Basis for Moral Behavior," *The Journal of Value Inquiry* 19 (1985): 43–53. Notable critiques include: Alasdair McIntyre, *After Virtue* (Notre Dame: University of Notre Dame Press, 1981); Charles Taylor, "Atomism," in his *Philosophy and the Human Sciences, Philosophical Perspectives*, vol. 2 (Cambridge: Cambridge University Press, 1985); H.J. McCloskey, "Rights," *Philosophical Quarterly* 15 (1966): 115–27; H.J. McCloskey, "Moral Rights and Animals," *Inquiry* 22 (1979): 23–25. Cf. also Tom Regan's critical response to McCloskey, "McCloskey on Why Animals Cannot Have Rights," *Philosophical Quarterly* 26 (1976): 251–257. And cf. Michael Zimmerman, "The Crisis of Natural Rights" and Joel Feinberg, "The Rights of Animals and Unborn Generations," *Philosophy and Environmental Crisis*, ed. William T. Blackstone (Athens: The University of Georgia Press, 1974), pp. 43–68. Also cf. Kenneth E. Goodpaster, "On Being Morally Considerable," *Journal of Philosophy* 75 (1978): 308–25. Goodpaster maintains that Feinberg mistakenly restricts rights to individual animals, and insists that entire bioregions can be said to have an interest of their own, too. Cf. also Christopher D. Stone, *Should Trees Have Standing? Toward Legal Rights for Natural Objects*, along with John Rodman's critique of Stone in Rodman's incisive essay, "The Liberation of Nature?" *Inquiry* 20 (1977): 83–131.

For a critique of the "natural rights" approach to environmental ethics, cf. J. Baird Callicott's review of Tom Regan, *The Case for Animal Rights, Environmental Ethics* 7 (1985): 365–72. Finally, cf. Kenneth E. Goodpaster, "From Egoism to Environmentalism," in *Ethics and the Problems of the 21st Century*, ed. Kenneth E. Goodpaster and K.M. Sayre (Notre Dame: Notre Dame University Press, 1979).

20. John Locke, *An Essay Concerning the True Origin, Extent and End of Civil Government*, in Burtt, *The English Philosophers from Hobbes to Mill* (New York: Modern Library) pp. 419–420: "For it is labor indeed that puts the difference of value on everything.... Nature and the earth furnished only the almost worthless materials as in themselves." Cf. Lorenne M.G. Clark, "Women and Locke: Who Owns the Apples in the Garden of Eden?" in *The Sexism of Social and Political Theory: Women and Reproduction from Plato to Nietzsche*, ed. Lorenne M.G. Clark and Lynda Lange (Toronto: The University of Toronto Press, 1979). Ellen Messer-Davidow (personal communication) has argued that "Locke is not guilty of dualism ... because he does not divide beings into two categories. Instead of only humans and animals or only males and females, he recognizes several categories: animals, women, men, children, various socioeconomic classes, and people who do not have full use of their faculties (his terms: idiots, etc.). In other words, ... Locke does not construct living beings in rigid dichotomies. He distinguishes not two but several categories."

21. On the Marxist treatment of nature, cf. Michael E. Zimmerman, "Marx and Heidegger on the Technological Domination of Nature," *Philosophy Today* 23 (1979): 99–112; Kostas Axelos, *Alienation, Praxis, and Techne in the Thought of Karl Marx*, trans. Ronald Bruzina (Austin: University of Texas Press, 1976); Alfred Schmidt, *The Concept of Nature in Marx*, trans. Ben Fowkes (London: New Left Books, 1971); Albrecht Wellmer, *The Critical Theory of Society*, trans. John Cumming (New York: Herder & Herder, 1971).

22. Naomi Scheman, "Individualism and the Objects of Psychology," in *Discovering Reality: Feminist Perspectives on Epistemology, Metaphysics, Methodology, and Philosophy of Science*, ed. Sandra Harding and Merrill Hintilcka (Boston: D. Reidel Publishing Company, 1983), p. 234.

23. Cf. Ruth Hubbard, "Have Only Men Evolved?" in *Discovering Reality*. For a nonfeminist criticism of the circularity of Darwin's rea-

soning about "natural" competition, cf. Jeremy Rifkin, *Algeny* (New York: Penguin Books, 1984).

24. Michael Gross and Mary Beth Averill, "Evolution and Patriarchal Myths of Scarcity and Competition," *Discovering Reality*, p. 72. Cf. Jane Flax, "Political Philosophy and the Patriarchal Unconscious," in *Discovering Reality*, for an insightful account of how Hobbes' competitive view of human nature is rooted in a masculinist philosophical anthropology.

25. On the antagonism between family and society, personal and public relationships, cf. Carole Pateman, "'The Disorder of Women': Women, Love, and the Sense of Justice," *Ethics* 91 (1980): 20–34.

26. Naomi Scheman, "Individualism and the Objects of Psychology," in *Discovering Reality*.

27. On object-relations theory, cf. Nancy Chodorow, *The Reproduction of Mothering* (Berkeley: University of California Press, 1978); Chodorow, "Family Structure and the Feminine Personality," in *Woman, Culture, and Society*; Dorothy Dinnerstein, *The Mermaid and the Minotaur: Sexual Arrangements and the Human Malaise* (New York: Harper & Row, 1976); Jane Flax, "Political Philosophy and the Political Unconscious." For a critical exchange on object-relations theory, cf. Judith Lorber, Rose Laub Coser, Alice S. Rossi, and Nancy Chodorow, "On the Reproduction of Mothering: A Methodological Debate," *Signs* 6 (1981): 482–514.

28. Carol Gilligan, *In a Different Voice: Psychological Theory and Women's Development* (Cambridge: Harvard University Press, 1982). For an appraisal of the moral development approach to ethical issues, cf. the essays in the special issue of *Ethics*, vol. 92 (April 1982).

29. Cf. John Rodman, "Animal Justice: The Counter-Revolution in Natural Rights and Law," *Inquiry* 22 (1979): 3–22.

30. Cf. John P. Briggs and F. David Peat, *Looking Glass Universe: The Emerging Science of Wholeness* (New York: Simon and Schuster, 1984); Nick Herbert, *Quantum Reality: Beyond the New Physics* (Garden City, NY: Anchor/Doubleday, 1985); John Briggin, *In Search of Shroedinger's Cat: Quantum Physics and Reality* (New York: Bantam Books, 1984); Paul Davies, *Other Worlds: Space, Superspace and the Quantum Universe* (New York: Simon and Schuster, 1980); Fritjof Capra, *The Tao of Physics* (New York: Bantam Books, 1982); Roger S. Jones, *Physics of Metaphor* (New York: New American Library, 1982); Charles

Birch and John B. Cobb, Jr., *The Liberation of Life* (Cambridge: Cambridge University Press, 1981). For a feminist approach to the new physics, cf. Robin Morgan, *The Anatomy of Freedom: Feminism, Physics, and Global Politics* (Garden City, NY: Anchor/Doubleday, 1982), especially the chapter called "The New Physics of Meta-Politics," and Marti Kheel, "The Liberation of Nature: A Circular Affair." For a deep ecological approach to the new physics, cf. J. Baird Callicott, "Intrinsic Value, Quantum Theory, and Environmental Ethics," *Environmental Ethics* 7 (1985): 257–75.

31. Cf. A.R. Peacocke, *Creation and the World of Science* (Oxford: Oxford University Press, 1979).

32. On the importance of feeling as a mode of disclosure in such realms as physical science, cf. Evelyn Fox Keller, *A Feeling for the Organism: The Life and Work of Barbara McClintock* (New York: W.H. Freeman and Company, 1983); *Reflections on Gender and Science* (New Haven: Yale University Press, 1985); "Feminism and Science," in *The Signs Reader: Women, Gender, and Scholarship*, ed. Elizabeth Abel and Emily K. Abel (Chicago: University of Chicago Press, 1983); "Gender and Science," *Discovering Reality*; "Women, Science, and Popular Mythology," *Machina ex Dea*.

33. There have been male philosophers who have emphasized the importance of feeling and sentiment in moral theory. David Hume is the most important example. Concerning the value of Hume's thought for environmental ethics, cf. J. Baird Callicott, "Elements of an Environmental Ethic," *Environmental Ethics* 1 (1979): 71–81; Callicott, "Hume's Is/Ought Dichotomy and the Relation of Ecology of Leopold's Land Ethic," *Environmental Ethics* 4 (1982): 163–74.

34. On this shift in the feminist movement, cf. Eisenstein, *Contemporary Feminist Thought*, and Eisenstein and Jardine, eds., *The Future of Difference*.

35. My thanks to Kathryn Carter for conversations that helped to clarify these issues.

36. Carole Pateman, "'The Disorder of Women': Women, Love, and the Sense of Justice," *Ethics* 91 (1980): 33.

37. Ibid.

38. Flax, "Political Philosophy and the Patriarchal Unconscious," *Discovering Reality*.

39. Kheel, "The Liberation of Nature," p. 140.

40. Salleh, "Deeper than Deep Ecology." The essays she criticizes are Arne Naess, "The Shallow and the Deep, Long-Range Ecology Movement," and Bill Devall, "The Deep Ecology Movement."
41. Salleh, "Deeper than Deep Ecology," p. 311.
42. Ibid.

14

Making Peace with Nature: Why Ecology Needs Feminism

Patsy Hallen

The 'control of nature' is a phrase conceived in arrogance, born of the Neanderthal age of biology and philosophy, when it was supposed that nature exists for the convenience of man. The concepts and practices of applied entomology for the most part date from that Stone Age of science. It is our alarming misfortune that so primitive a science has armed itself with the most modern and terrible weapons, and that in turning them against the insects, it has also turned them against the earth.

Rachel Carson, *Silent Spring*

This paper aspires to illuminate pathways of making peace with Nature. This is not an easy task since everywhere we are so much at war with the Earth, from DDT to deforestation, from acid rain to radioactive fallout.[1]

A Multi-Dimensional Thesis

My central thesis is that if we are to make peace with Nature ecology needs to be transformed by the knowledges[2] of feminism. This is a complex and multi-dimensional thesis involving our individual and collective psyche (the psycho-sexual realm), our ways of knowledge (science), and our ways of being (society and the environment). To support this thesis I will draw from philosophical, psychological, sociological and historical sources not in a linear, combative way (statement of main thesis, arguments for main thesis, defeat of objections to main thesis, and so on) but in a spiral, processive way. I do not propose to argue. "Any kind of polemics," writes Heidegger, "fails from the outset to assume the attitude of thinking."[3] I do not propose to be polemical. What I hope to do is to evoke and share a vision.[4] This invitation-to-look approach is a deliberate strategy designed to stand as a testimony to the complexity and the vitality of the deceptively simple claim: *ecology needs feminism.*

Science Must Be Mediated by Feminist Knowledge

Feminist scholarship has shown how modern science is basically a masculine endeavor. As a result, I will suggest, science is one-sided, and potentially life-threatening. We will not overcome this one-sidedness, I will maintain, until our scientific understanding of the living world is mediated by the content of feminist epistemologies— the sensual, the relational, the intimate. Unless our science is grounded in the "caring labor" articulated by feminism,[5] science will continue to be, as Rachel Carson suggests, "against the earth."

Furthermore, I will maintain that these feminist perspectives (which have their roots in traditional thinkers from Plato through Kierkegaard to Polanyi[6]) have been systematically ignored because of our individual and collective psychotic need to dominate. Ecology's wholesomeness, I will posit, is linked in a profound way to the absence of a need to dominate. This claim, then, necessitates an investigation of the psycho-sexual dimension to uproot the causes of domination.

Thus a related thesis which supports my main thesis is this: that the current ecological crisis has psycho-sexual roots. The ecological crisis has many other causes as well (economic and political), but we have underplayed, even ignored the psychosexual causes until feminist theory invited us to investigate this dimension.[7]

Why Science Needs Feminism

Following feminist writers I will maintain that sexism is the expression of a basic psychology of domination and repression. Ecological imbalance is, in part, due to our mistaken[8] belief that we can successfully dominate Nature. So sexism (mind and body pollution) is fundamentally linked to ecological destructiveness (environmental pollution).

I will suggest that ecology as a *science* needs feminism to balance a myopic, mechanical worldview which has fundamentally influenced scientific development; ecology as a *life science* needs feminism to reveal how patriarchal thinking contributes to environmental destruction; ecology as a practice needs feminism to ensure that shallow ecology is transformed into a deep ecological perspective. The shallow/deep distinction was drawn by Norwegian philosopher Arne Naess, to contrast reform environmentalism with deep ecology. Reform environmentalism aims to manage the environment based on its use-value, while deep ecology seeks to help us see and feel ourselves as intimately interrelated to an intrinsically valuable Nature, so that when we harm Nature we diminish ourselves.[9]

I will posit that one of the pathways to making peace with Nature is through feminism. Feminism is helpful to ecology as a science, as a life science, and to its practitioners because the multiplicity of differences within feminism offers new ways of experiencing and understanding the world.

Despite the main theoretical differences between feminists (liberal, Marxist, socialist, post-structuralist, radical), feminists adhere to a few basic tenets, as outlined by Marilyn French:[10] that the two sexes are at least equal in all significant ways and that this equality must be publicly recognized; that the qualities traditionally associated with women (nurturing, receptivity) are at least equal in value to those traditionally associated with men (self-assertion, power-seeking) and that this equality must be publicly recognized, not ignored or viewed as irrelevant; finally that the personal is the political, the bedroom is as relevant as the boardroom.

What I hope to show is that each of these tenets of feminism can help to enlarge our scientific understanding of the environment. The first tenet will ensure that more women will participate in the scientific community.

It is crucial that there be more women in science. For as Harding points out, science is the model in our culture of a masculine activity. To quote Harding, "women have been more systematically excluded from doing serious science than from performing any other social activity, except perhaps frontline warfare."[11] Hence, to redress the scarcity of women in the peculiarly masculine occupation of scientist is vital.

But this, while necessary, is not sufficient. We do not just need more women in science; the whole nature of the system, the dominant ideology of science, needs to be challenged.

As the researchers of women's struggles to enter science show,[12] even if horizontal segregation (science as an exclusively male domain) is to some degree overcome, vertical segregation (women scientists confined to low status positions) has yet to be eliminated. We not only need more women scientists, we need women to be equally recognized practitioners of science. As Hilary Rose points out,[13] the majority of people actually practicing science—technicians—are women, but their work is marginalized, trivialized, made invisible, and this is so even when the content of the scientific work done by women is objectively indistinguishable from men's work. This indicates that other, deeper factors are at work. Are masculine gender identities so fragile that they cannot afford to have women as equals to men in science, Harding asks?[14] Until the 'emotional labor' of childcare and housework is seen as desirable for men, the 'intellectual labor' of science and public life will not be perceived as desirable for women. To transform science will require revolutionary change in the social relations between the sexes.

Even if there are more equally recognized female practitioners of science, science's relations to society need to be altered. For women do not want to become 'just like men' in an enterprise that, in the USA for example, devotes 72 percent of its federal funding for scientific research and development to defense.[15] Furthermore, women should not want to become puzzle-solvers in vast industrialized empires devoted to material accumulation, social control and exploitation, activities leading to species extermination and environmental degradation. It is not only bad science (hastily generalized science with inappropriate data bases), but 'normal' science that needs to be criticized since, as the critics of modern science show,[16] so much of it facilitates militarism, ecological disasters and desperate poverty.

By the very same token, we do not just need more ecologists; for ecology can be exclusively reductionist, 'understanding' life in terms of nonlife, and so failing to see living things in process, in relation and in context, in terms of dynamic energy patterns. In addition, all-too-often environmental impact statements are fronts, hoaxes. As Neil Evernden points out,[17] universities willingly disgorge troops of environmental scientists and managerial environmentalists, and while they appear to be tools of environmental defense, they turn out— lo and behold!—to serve the interests of the developer. So for truly emancipatory knowledge-seeking, the first tenet of feminism, more women ecologists, is not sufficient.

We need the second tenet of feminism whereby the so-called 'feminine' qualities will be cherished and female experience can be incorporated into explanations. This will help to offset what Evelyn Fox Keller calls the "hegemony in science,"[18] the 'masculine', virile, domineering nature of science, and to reclaim science as a human, not a masculine project, and so to, in Keller's words, "transform the very possibility of creative vision."[19]

This key rationale will be further substantiated in the course of this paper. For now, suffice it to say that, as Keller points out,[20] despite the wide diversity in the practice of science, there is a monolithic ideology of detachment and domination which crystalized in the 17th century, and this ideology has deeply influenced the selection of goals, the methods, values and explanations that operate in contemporary science.

While this view draws our attention to how so-called 'feminine' thinking—intuitive and relational thinking—has advanced scientific comprehension, the perspective of valuing feminine qualities is also not sufficient. It is vital to point out the importance of pluralism in the intellectual pursuit of knowledge, but until we understand what motivates the exclusion of these non-macho elements in our scientific story-telling, we will not understand why some ideas gain legitimacy and others do not. So it is necessary to investigate the psychological profile of masculinity: a third level of analysis—the psychosexual.

This third tenet of feminism—the personal is the political or the sexual is the scientific—announces a three-storied method.

Firstly, as feminists point out, the public sphere depends upon the private resource base. But this private resource base is often not acknowledged, or if it is, it is trivialized. Feminism insists that we ask why it is trivialized and that we come to terms with the importance of this private dimension.

Secondly, feminist analysis shows us that who we are (in the bedroom) shapes our politics (in the boardroom). So feminism insists that our sexuality, our sense of self, be articulated in any intellectual endeavor. We must 'begin at the start'. We must start from the personal since it is so formative.

Thirdly, feminism seeks to enlarge our understanding of Nature and the world not just by including feminine experience in explanations, but also by insisting on including those domains of human experience that have been relegated to women: namely the personal, the emotional, the sexual. An example of this method is Brian Easlea's book, *Fathering the Unthinkable*,[21] which analyses a scientific achievement—the building of the bomb—in psychosexual terms. I [agree] ... that including the personal, emotional, sexual dimensions of experience in our explanations will and does make us better scientists, because they alert us to our signatures and in the process help to enlarge our sense of what is possible.

But such theories calling for the integration of the personal and the scientific will be "fated to remain mere intellectual curiosities—like the ancient Greek ideas about atoms,"[22] until we understand the unparalleled importance of our sexuality and the relation between sexual repression and domination.[23] Just as ontogeny (the development of an individual) recapitulates phylogeny (the evolution of the species)—for example, a human embryo mirrors our reptilian and amphibian past in its life history—so too cosmogeny (world-building) recapitulates erogeny (sexual building).[24] Until we understand these principles, we will not achieve psychic health or scientific wholeness.

A Relational View of Humans and Nature

Feminism acknowledges the central importance of the erotic, the private, the personal, and as a result it speaks "in a different voice," to borrow Carol Gilligan's phrase.[25] It offers us a mode of thought at variance with the dominant ideology. Its central ontological category is not substance, but relation. This category transcends mere relativism, goes beyond and beneath divisive dualisms and dichotomies and insists on the scientific importance of the personal. To paraphrase Hilary Rose, [a feminist epistemology] transcends dichotomies, insists on the scientific validity of the subjective, on the need to unite cognitive and affective domains; it emphasizes holism, harmony and complexity rather than reductionism, domination and linearity.[26]

Rose argues that this relational view is easier for women to attain because of their labor. Women's labor is more "in touch with necessity."[27] It is caring labor reflecting the unification of mental, manual and emotional activities ("hand, brain and heart") characteristic of women's work. As such it is more complete and therefore "truer knowledge."[28]

In addition, I surmise this relational view might be easier for women to attain because of their gender identity. Psychologists and social theorists who have studied gender identity (as distinct from sexual identity) tell us that femininity is defined through attachment and connection, while masculinity is defined by independence and separation.[29] The male gender is threatened by intimacy; the female gender is threatened by separation. So gender-sensitive theories show us how to better relate to our environment, show us how to more fruitfully connect with Nature.

It is no accident that ecology, which etymologically means 'a study of the household', needs the experience of women. To see the Earth as a life-sustaining home is a vision of ecology which, I believe, is accessible to women.

Scientifically Revolutionary Consciousness

As a result of women's caring labor and women's gender identities, I propose to you that feminist consciousness is a scientifically revolutionary consciousness. I see feminism and ecology as sharing the same perspective, which represents a new (yet very old) way of seeing, a way of making peace with (rather than war on) Nature. Such a perspective is holistic: everything is connected to everything else and each aspect is defined by and dependent upon the whole, the total context. Life is interconnected and interdependent: we are not above Nature, we are an intimate part of it.

We have been blinded to holism by the gender ideology of science, by our arrogance, by our false sense of superiority, by our fragile identities rendered vulnerable by projecting our inner tensions outward onto 'the second sex' rather than working them through, by our incomplete notions of survival of the fittest with its resultant stress on competitiveness. Even David Attenborough's superb television series, "Life on Earth," was biased in this way. Its scenes of intense competition far outweighed its scenes of cooperation in the evolutionary story. As John Livingston points out, most wildlife biologists carry around an extraordinary load of assumptions: concepts like dominance, aggression and competition—a "long roster of market-place concepts applied to nature."[30] In addition to this excess baggage, we have been blinded to holism by the spatial metaphor with which we grew up, God a ruling class male above, then man with woman and Nature (mythed as female) below. This hierarchical picture has dominated Western thought for hundreds of years and it is still very much alive despite the 20th-century revolutions in scientific thinking. I propose that we unconsciously imbibe this worldview, we smuggle it into our conscious mind and let it structure our thoughts.

Both ecology and feminism challenge this hierarchical picture. Both reject the divisive dualism that this worldview implies: heaven-earth, male-female, mind-matter, reason-emotion, objective-subjective. To develop the masculine and the feminine in each of us, to become whole people, to be able to see objects as subjects, to see Nature as a thou, to experience the Earth as a live presence, to feel Nature as part of ourselves—the visions of ecology and feminism dovetail. Cooperation is stressed (not competition), understanding is vital (not power), appreciation is important (not domination). The watchwords are solidarity and sharing, not rivalry and ruling. As Mary O'Brien puts it, what is important is "reciprocal intimacy and not conquest."[31] Both ecology and feminism share a non-hierarchical, egalitarian perspective. Both participate in a common philosophy whereby process and participation are primary. Process philosophy is a deep and complex topic with its sources in the philosophies of Alfred North Whitehead and Henri Bergson.[32] For our purposes it is sufficient to say that feminism and ecology both stress creative activity over inert matter, dynamic order over static laws, partial autonomy over determinism, relation over substance, objects as subjects over subjects as objects. Finally, for both ecology and feminism, as Carolyn Merchant points out,[33] there is no free lunch. This is one of the four laws of ecology articulated by Barry Commoner.[34] We may think that we are getting a cheap digital watch at $4.99, but the real cost must include the Formosan women who go blind producing the liquid crystals. There is no free lunch. To produce organized matter, energy in the form of work is needed. Reciprocity and cooperation oiling the feedback loops and closing the energy circuits is what is needed, not free lunches.

Whereas patriarchal society needs an inferior 'other' in terms of which to define itself—an insight so clearly articulated by Simone de Beauvoir—(whether that other be 'woman' or 'African' or 'black')[35]—a society transformed by feminist consciousness would oppose such hegemony. Feminism argues for a plurality of discourses and for a vision of wholeness based not on divisive dichotomies or defensive competitiveness but on a genuine appreciation of the difference.

The field of primatology has attracted a disproportionate number (compared to other related fields) of woman researchers, and it is here that we can witness feminist consciousness in action: the explosion of the old models of male-dominated power structure, the realization of the importance of co-operation and the undermining of this category of 'other.'[36] Likewise as Livingston points out, a study of short-grass prairie songbirds in New Mexico came to the (surprised) conclusion that "competition is not the ubiquitous force that many ecologists have believed."[37]

Both the female primatologists and John Wiens' papers on the short-grass songbirds challenge the notions of competition, dominance and the concept of 'the other'. They show us how in Nature the self/other distinction is often inappropriate. Success for a species depends upon individual bonding with the environment—an extended consciousness of self that transcends an isolated self. This "participating consciousness" is crucial for the survival of many species, and it is crucial for human survival. And the perception in sensitive science that the environment or the non-self is not 'other' is music to feminist ears.

Reciprocal Intimacy in Science

Let me share with you a striking example of how science might be different, of how science might look, when it is based on an attitude of reciprocity and cooperation rather than a competitive 'other'. This attitude of "reciprocal intimacy" in science is illustrated by Barbara McClintock, who in 1983 won a Nobel Prize for her work in biology in the 1940s. Evelyn Fox Keller has written a moving biography of McClintock called *A Feeling for the Organism*[38] and in it she asks:

> What is it in an individual scientist's relation to Nature that facilitates the kind of seeing that eventually leads to productive discourse? What enabled McClintock to see further and deeper into the mysteries of genetics than her colleagues?

Her answer, Keller tells us, is simple:

> Over and over again McClintock tells us one must have the time to look, the patience to hear what the material has to say to you, the openness to let it come to you. Above all one must have a feeling for the organism. And McClintock goes on to say: 'No two plants are exactly alike. They are all different. And as a consequence you have to know that difference.' She explains: 'I start with a seedling and I don't want to leave it. I don't feel I really know the story if I don't watch the plant all the way along. So I know every plant in the field. I know them intimately and I find it a great pleasure to know them.'

McClintock calls herself "a mystic in science," this woman whose work has been recognized with a Nobel Prize, and she says that her aim is to "embrace the world," according to Evelyn Keller, "in its very being, through reason, and beyond." Now to embrace the world is something very different, of course, than the desire to conquer it.

Barbara McClintock's articulated desire to respect and embrace the world stands in stark contrast to Francis Bacon's expressed desire to "put nature on the rack and torture her"[39] so that Nature—the ultimate 'other'—will reveal her secrets. Hence, there is within science a very different tradition than that of domination, manipulation and control exemplified by one of the founding fathers of modern science, Francis Bacon. And it is this alternative tradition of science, represented by Barbara McClintock's love of plants, that we need to understand and emulate, if we are going to have an ecologically sound and sustainable society.

.McClintock shows us how to open our other eye, how to overcome our long history of one-sidedness. To McClintock science is not premised on detachment, on domination, on a division between subject and object, between self and 'other'. In a more recent book, *Reflections on Gender and Science,* Keller states that McClintock's love of Nature allowed for "intimacy without the annihilation of difference."[40] As Keller says, "division is relinquished without generating chaos."[41] A vivid illustration of this love, this form of attention to things, comes from McClintock's own account of a breakthrough in one particularly recalcitrant piece of analysis. McClintock describes the state of mind accompanying the crucial shift in orientation that enabled her to identify chromosomes she had earlier not been able to distinguish.

> I found the more I worked with them, the bigger and bigger the chromosomes got and when I was really working with them, I wasn't outside. I was part of the [system] ... it surprised me because I actually felt as if I was right down there and these were my friends.... As you look at these things they become part of you....[42]

This account of matter is a far cry from the dead inert matter of the mechanical model. As Keller points out, McClintock allows us to see the profound kinship between us and Nature. She encourages us to witness the astonishing diversity and unimaginable resourcefulness of the natural order. She inspires a real feeling for the organism. McClintock sees Nature not as blind, simple and obedient, but as self-generating, complex and resourceful. Nature, for

McClintock, is not more complex than we know, but more complex than we can know. Nature is ingenious. "Anything you can think of, you will find," declares McClintock.[43] Hence, for McClintock the goal of science is not the power to manipulate, but empowerment, the power to understand, the power to appreciate, the power to humble.

Different Voices Showing Our Ideological Biases

There are real limitations in contemporary science, and feminist writers like Keller, and scientists such as McClintock, are helping us to see the ideological biases of science, and so to transcend them. Keller notes that McClintock is not a feminist scientist,[44] since McClintock's vision was of a science not based on sex or gender. But here Keller misses the point: feminism wishes just that. Feminism seeks to transcend the dividing dichotomies between masculine and feminine and to found a new science based on McClintock's vision. What is encouraging is that there is a shift in emphasis in numerous fields which gives weight to the argument developed here.

Nel Noddings revolutionized ethics through her book, *Caring*,[45] by showing how morality (and ethics) was dominated by certain values and concerns, by talk of principles and justification, rather than by talk of caring and receptivity, relatedness and responsiveness. In her Introduction she says:

> One might say that ethics has been discussed largely in the language of the father: in principles and propositions, in terms such as justification, fairness and justice. The mother's voice has been largely silent. Human caring and the memory of caring and being cared for, which I shall argue form the foundation of ethical behavior, have not received attention, except as outcomes of ethical behavior.[46]

If care is the foundation of our understanding we will include both categories 'mother' and 'father' as Noddings does to embrace a new way of relating to Nature.

Likewise, Erazim Kohák in his book *The Embers and the Stars*[47] states that the aim of the book is to shift the burden of philosophizing from the making of arguments to the aiding of vision. Not cunningly devised theories but encountering the wonder of being, this is the objective.

We need these different voices. We need them urgently. Theories can be very influential but to help them be so, as Herbert Marcuse argued in *Eros and Civilization*,[48] they must move from the surface and delve into the deepest biological layers of human energies, the well-springs of human action. Otherwise our theories of political, economic, social or ecological change are rootless; we are picking up litter, rather than attacking the production of unstable waste in the first place. And this delving into the biological layers is precisely what feminist theory is doing.

Science and the Interests of Power

All modern, scientific thinking is at bottom power thinking, that is to say, the fundamental human impulse to which it appeals is the love of power, or, to express the matter in other terms, the desire to be the cause of as many and as large effects as possible.[49]

Ever since Bacon announced that knowledge is power, science has been dominated by a will to power and so has been domineering, as Bertrand Russell argues. Bacon talks about how the new science will be a rebirth. He jeers at the Greeks, who he calls "mere boys," and calls for a masculine birth of time, a rebirth without indebtedness to women.[50] Through this rebirth, the scientific imagery goes, one can conquer the universe. And what does conquering the universe involve? The "death of nature," as Carolyn Merchant points out.[51] At the same time as Bacon, Descartes declared Mother Nature dead—mere matter in motion without sentience, without consciousness. Animals are mere machines, in the Cartesian framework.

Hence, in the 17th century we have a cognitive attack on Mother Nature through the writings of such influential thinkers as Bacon and Descartes. Coupled with this assault we also have a physical attack on women. In the 16th and 17th centuries scores of thousands of women were hunted and killed for witchcraft. This link is, Easlea argues,[52] no accident. It spells a deep antagonism to women which men cope with by being violent towards Nature. So anti-feminist sentiment feeds ecological disaster, from the testing of atomic bombs to using animals as tools of trivial research projects.

From Mother Earth to an Inert Machine

The patriarchy's envy and fear of women and the Earth was transformed symbolically into a mechanistic worldview. As Merchant illustrates,[53] the Earth was no longer regarded as a nurturing mother to be cherished, but as a machine to be manipulated and exploited. The transformation from Earth mother to barren inert matter had been accomplished through the scientific revolution of the 17th century and men hoped to become, through the medium of science, "masters and possessors of nature."[54]

This mechanical worldview accelerated the exploitation of humans, both men and women, and the exploitation of natural resources. Mechanism saw Nature not as a live presence, but as a system of dead, inert particles moved by external forces. As such, it bears no moral self-examination. Nature is merely material for man's appropriation, a view which suited commercial capitalism.

Both feminism and ecology challenge mechanism's assumptions which make the manipulation and control of Nature seem possible and acceptable. The ontological assumption of mechanism is that matter is not living, interdependent fields of energy, but matter is composed of dead, discrete particles. The epistemological assumption is that knowledge and information can be abstracted without distortion from the natural world, because they are independent of context and value-free. The methodological assumption is that real-life problems can be successfully analyzed into parts that can be manipulated by mathematics. The moral assumption is that humans (especially white, upper-class males) are more valuable than the rest of Nature.

The mechanical paradigm, though outdated, persists. Francis Crick states: "The ultimate aim of the modern movement in biology is, in fact, to explain all biology in terms of physics and chemistry."[55] The project is to explain life (humans, brains) in terms of non-life (machines, computers). The commitment is that Nature is lifeless. And this allows us to perform gruesome experiments on living animals and to write them up "as if the experiments had been done on inert matter."[56]

Contrast two descriptions: Woldridge, a modern psychological researcher, convinces himself that when a monkey with an electric current passing through its "pain center" bites objects so hard as to wrench teeth from its jaw—then the monkey is really experiencing pain and not exhibiting "meaningless" or "automatic physical symptoms."[57] Compare this to a description of a chimpanzee being studied by Jane Goodall in the Tanzanian National Park of Gombe Stream. Once a chimpanzee grasped her hand "firmly and gently with his own before scurrying off into the forest." Goodall writes that "at that moment there was no need of any scientific knowledge to understand this communication of reassurance."[58]

Goodall's attitude to non-human life forms is, while superior, not a female prerogative. Martin Buber, for example, has an I-Thou relationship while looking into the eyes of his cat.[59] And people who live close to Nature see Nature not as an 'it', but as a 'thou'. Nature is for so-called 'primitive' people a live presence, each part of which is unique, and not satisfactorily describable by a universal law.[60] Hence, their relationship with a tree, for example, is emotional, direct, not fully able to be articulated. For many traditional peoples, "Nature is their body."[61]

> Communication at its best is called love; when it breaks down completely, we call it war. And it is a sort of war that is going on now between human beings and the earth. It's not that nature refuses to communicate with us, but that we no longer have a way to communicate with it. For millennia, primitives communicated with the earth and all its beings by means of rituals and festivals where all levels of the human were open to all levels of nature.[62]

We need to recapture this reality. We need, as A.N. Whitehead urges, to develop "a science based on an erotic sense of reality rather than an aggressive, dominating attitude towards reality."[63]

We need to relate the crisis of production (and overproduction) to the crisis of reproduction.[64] We will not solve our environmental crisis unless we allow the Earth to reproduce.

We need to appreciate reproductive values, such as a recognition of our own death. Freud did not consider aggression a basic psychological fact. Aggression for Freud was rather a secondary manifestation of a more fundamental instructural force, the death instinct.[65] Norman O. Brown, analyzing Freud in *Life Against Death*,[66] felt that we could learn to channel our aggressive behavior positively, if only we could learn to contain death within life, that is, if we were not "fugitives from our own death,"[67] in short, if we learned how to die with dignity.[68]

As Barbara McClintock reminds us, we need a new scientific basis whereby we recognize that each organism is unique and that it is an integral part of an ecosystem. We need a new psychological basis whereby we see, as Carl Jung reminds us,[69] that to heal is to make whole, as in wholesome, to reunify our split selves, to integrate the masculine and the feminine in each of us. We need a new epistemological basis whereby we realize, as Norman O. Brown reminds us,[70] that we must have carnal knowledge, a copulation of subject and object. We need a new ethical basis whereby we recognize, as the Aboriginal peoples show us, the intrinsic value of and our dependence upon the non-human aspects of Nature. Finally, we need a new ontological basis whereby we experience, as Hegel details,[71] that reality is a process and that the truth is just as much subject as substance—that the truth is the whole.

Why We Need Each Other

We need to recognize that the whole is more than the sum of the parts and that the parts take their meaning from the whole. Each part is defined by and dependent upon the total context. Isolation (as in a laboratory) distorts the truth because it distorts the whole. As it is valid to interpret the higher (life) in terms of the lower (non-life), so it is also valid to interpret the lower in terms of the higher. 'Aim' can be applied to cells; 'enjoyment' can be applied to gorillas. Reality is a complex and dynamic web of energy. Nature is alive and active and its parts are fundamentally interconnected by cyclical and developmental processes.

We need a reversal of mainstream, malestream values and the triumph of feminism, a revolution in economic priorities and a steady-state economy,[72] a peace force for a just and sustainable society,[73] a social force for voluntary simplicity,[74] and collective action for the ecological reconstruction of society.[75]

We need to overcome our dichotomies and to discover our deep sources, our springs, as Rachel Carson did.

We need each other.

Notes

1. In this paper I talk about women being more in touch with nurturing than men. This talk is problematic insofar as it might reinforce traditional stereotypes of women. The variability among men and women in a wide range of diverse traits is much greater than the difference between men and women. (See Marion Namerwirth, "Science Seen through a Feminist Prism," in Ruth Bleier, ed., *Feminist Approaches to Science*, New York: Pergamon Press, 1986, p. 38.) In this paper I also draw upon the distinction between sex and gender. This is a very useful distinction separating our biology from our socialization, our sex from our sexuality. But it is also a problematic distinction highlighted by the argument about the role of Nature vs. nurture. Gender is a fundamental organizing category, and as Sandra Harding points out, whether it applies at the individual level, at the structural level or at the symbolic level, gender is always asymmetrical, with the feminine defined (in Western culture since the Christian era) as inferior. (See Sandra Harding, *The Science Question in Feminism*, Ithaca: Cornell U. Press, 1986, esp. pp. 55–57.)

2. Michel Foucault, *A History of Sexuality*, Vol. I, New York: Random House, 1980.

3. Martin Heidegger, "What Calls for Thinking?" in D.F. Drell et al. (eds.), *Basic Writings*, New York: Harper and Row, 1977, p. 354.

4. Erazim Kohák's goal in his eloquent book *The Embers and the Stars: A Philosophical Inquiry into the Moral Sense of Nature*, Chicago: University of Chicago Press, 1984, p. xiii.

5. See, for example, Hilary Rose, "Beyond Masculinist Realities: A Feminist Epistemology for the Sciences," in Ruth Bleier (ed.), op cit.,

Feminist Approaches to Science pp. 55–76; or Hilary Rose, "Hand, Brain and Heart: A Feminist Epistemology for the Natural Sciences," *Signs: Journal of Women in Culture and Society* 9, No. 1, 1983.

6. Plato, *The Collected Dialogues Including the Letters*, edited by Edith Hamilton and Huntington Cairns, New York: Pantheon Books, Bollingen Series LXXI, 1961; Soren Kierkegaard, *Concluding Unscientific Postscript*, translated by Walter Lowrie, Princeton: Princeton University Press, 1956; Michael Polanyi, *Personal Knowledge*, New York: Harper Torchbooks, 1958.

7. For example Simone de Beauvoir's fruitful framework of analyzing women's status as 'the other'. See *The Second Sex*, trans. & ed. by H.M. Parchley, Harmondsworth, Middlesex, England: Penguin, 1975.

8. For instance, the treadmill situation of pesticide use in California. See Robert van den Bosch, *The Pesticide Conspiracy*, New York: Doubleday, 1978. Bosch documents how pesticides poison humans, other animals and birds, not pests. We have not successfully eliminated one species of pest, only their predators. Hence, we are using vast quantities of pesticides and still losing more crops to the resilient and genetically variable pests than ever before: a treadmill situation.

9. Arne Naess, "The Shallow and the Deep, Long-Range Ecology Movement," *Inquiry* 16 (1973), pp. 95–100; see also Ariel Kay Salleh, "Deeper than deep ecology: The eco-feminist connection," discussion papers, *Environmental Ethics*, Winter, 1984.

10. Marilyn French, *Beyond Power*, New York: Harper and Row, 1985.

11. Harding, *The Science Question in Feminism,* op cit., p. 31.

12. Michele Alrich "Women in Sciences," *Signs: Journal of Women in Culture and Society* 4, No. 1, 1978; Rita Arditti, "Women Drink Water: Men Drink Wine," in R. Arditti, P. Brennan and S. Cavrak (eds.) *Science and Liberation*, Boston: Solithend Press, 1980; Alice Rossi, "Why So Few?," *Science*, 148, (1965): 1196; Margaret Rossiter, *Women Scientists in America: Struggles and Strategies to 1940*, Baltimore, MD: Johns Hopkins University Press, 1982.

13. Rose, "Beyond Masculine Realities," pp. 60–62.

14. Harding, *The Science Question in Feminism,* op cit., p. 64.

15. *Science*, 9 April 1985.

16. Hilary and Steven Rose (eds.), *Ideology of/in the Natural Sciences*, Cambridge, MA: Schenkman, 1976.

17. Neil Evernden (ed.), *The Paradox of Environmentalism*, symposium proceedings, The Center for Environmental Studies, Toronto: York University, 1984, p. 9.

18. Evelyn Fox Keller, *Reflections on Gender and Science*, New Haven, Yale University Press, 1985, p. 4.

19. Ibid., p. 178.

20. Ibid., p. 134.

21. Brian Easlea, *Fathering the Unthinkable: Masculinity, Scientists and the Nuclear Arms Race*, London: Pluto Press, 1983.

22. Harding, *The Science Question in Feminism*, op cit., p. 146.

23. Normon O. Brown, *Life Against Death*, New York: Random House, Vintage Books, 1964.

24. These ideas were born for me in listening to the lectures of Professor John Raser, a Foundation Professor of Murdoch University, Perth, Western Australia.

25. Carol Gilligan, *In a Different Voice*, Cambridge: Harvard University Press, 1982. The difference between relative and relational has been argued succinctly by feminists. See especially, Jane Flax, "Why Epistemology Matters," *Journal of Politics*, Vol. 27, 1981.

26. Hilary Rose, "Beyond Masculinist Realities," p. 72.

27. Marilyn French, *The Women's Room*, quoted by Rose, Ibid., p. 68.

28. Rose, Ibid., p. 72.

29. Nancy Chodorow, Dorothy Dinnerstein, Jessica Benjamin, Jane Flax and Ann Oakley, among others.

30. John Livingston, "The Dilemma of the Deep Ecologist," in Neil Evernden (ed.), *The Paradox of Environmentalism*, p. 63.

31. Mary O'Brien, *The Politics of Reproduction*, Boston: Routledge and Kegan Paul, 1983.

32. Alfred North Whitehead, *Process and Reality*, corrected edition edited by David Ray Griffin and Donald W. Sherburne, New York: Collier, Macmillan, 1978; Henri Bergson, *Creative Evolution*, trans. by Arthur Mitchell, New York: Random House, The Modern Library, 1944.

33. Carolyn Merchant, "Feminism and Ecology," Appendix B, pp. 229–231 in Bill Devall and George Sessions, *Deep Ecology: Living as if Nature Mattered*, Salt Lake City, Peregrine Smith Books, 1985.

34. Barry Commoner, *The Closing Circle: Nature, Man and Technology*, New York: Alfred A. Knopf, Inc., 1972, p. 41.

35. Harding, *The Science Question in Feminism*, Chap. 7, pp. 163–196.
36. Donna Haraway, "Primatology is Politics by Other Means" in Ruth Bleier, *Feminist Approaches to Science*, pp. 77–118.
37. Livingston, "The Dilemma of the Deep Ecologist," p. 67.
38. Evelyn Fox Keller, *A Feeling for the Organism: The Life and Work of Barbara McClintock*, New York: W.H. Freeman, 1983. All quotes in this paragraph are from pp. 99–102.
39. Francis Bacon, "Thoughts and Conclusions" in Benjamin Farrington, *The Philosophy of Francis Bacon*, Liverpool: Liverpool University Press, 1970, p. 92; see also F. Bacon, *Essays*, Ward, Lock and Bourden, 1894.
40. Keller, *Reflections on Gender and Science*, p. 164.
41. Ibid., p. 165.
42. Ibid.
43. Ibid., p. 162.
44. Evelyn Fox Keller, "Women, Science and Popular Mythology" in *Machina ex Dea: Feminist Perspectives on Technology* (cited in a personal correspondence).
45. Nel Noddings, *Caring: A Feminine Approach to Ethics and Moral Education*, Berkeley: University of California Press, 1984.
46. Ibid., Introduction, p. 1.
47. Kohák, *The Embers and the Stars*.
48. Herbert Marcuse, *Eros and Civilization*, New York: Vintage Press, 1962.
49. Bertrand Russell, *The Scientific Outlook*, London: Allen and Unwin, 1949.
50. Francis Bacon, "The Masculine Birth of Time and the Great Insaturation of the Dominion of Man over the Universe" in B. Farrington, *The Philosophy of Francis Bacon*.
51. Carolyn Merchant, *The Death of Nature: Women, Ecology and the Scientific Revolution*, San Francisco: Harper and Row, 1980.
52. Brian Easlea, *Witch-hunting, Magic and the New Philosophy: An Introduction to Debates of the Scientific Revolution 1450–1750*, Brighton, Sussex: Harvester Press, 1980, p. 10.
53. Merchant, *The Death of Nature*.
54. Bacon, "The Masculine Birth of Time," p. 28.
55. Francis Crick, *Of Molecules and Men*, New York: Pan Books, 1968.

56. Brian Easlea, *Liberation and the Aims of Science*, London: Chatto and Windus Ltd., 1973, p. 263.

57. D.E. Woldridge, *The Machinery of the Brain*, New York: McGraw Hill, 1963, p. 130.

58. Jane van Lawick-Goodall, *In the Shadow of Man*, Glasgow: William Collins, Fontana Books, 1974, p. 184.

59. Martin Buber, *I and Thou*, New York: Clark, 1970.

60. Henry Frankfort, et al., *Before Philosophy*, Hammondsworth, Middlesex: Penguin Books, 1959, pp. 12–13.

61. Karl Marx, "1844 Manuscripts," in *The Early Texts*, ed. D. McLellan, Oxford: Oxford University Press, 1962, p. 117.

62. Dolores LaChapelle, *Earth Wisdom*, Silverton, CO.: Guild of Tutor Press, 1978, p. 96.

63. Alfred North Whitehead, *Modes of Thought*, New York: Capricorn Books, 1958, p. 202.

64. O'Brien, *The Politics of Reproduction*.

65. Paul Robinson, *The Sexual Radicals*, London: Paladin, 1972, p. 169.

66. Norman O. Brown, *Life Against Death*, p. 110 ff.

67. Martin Heidegger, *Being and Time*, trans. by John Macquarrie and Edward Robinson, New York: Harper and Row, 1962, Part I, Division 2, H235 ff, pp. 279–311.

68. Elizabeth Kübler-Ross, *On Death and Dying*, London: Tavistock Publications, 1969.

69. Carl G. Jung, *The Undiscovered Self*, trans. by R.F.C. Hull, New York: The New American Library, Mentor Books, 1959, p. 119 ff.

70. Norman O. Brown, *Love's Body*, New York: Random House, 1966, p. 17.

71. G.W.P. Hegel, *The Phenomenology of Mind*, trans. by J.B. Baillie, London: George Allan and Unwin Ltd., 1964, Preface.

72. Herman Daly, *Steady State Economics*, San Francisco: W.H Freeman, 1977.

73. A phrase coined by the World Council of Churches; see P. Abrecht, *Faith, Science and the Future*, Geneva: World Council of Churches, 1978.

74. K.S. Shrader-Frechette, "Voluntary Simplicity and the Duty to Limit Consumption" in *Environmental Ethics*, ed. K.S. Shrader-Frechette, Pacific Grove, CA: The Boxwood Press, 1981, pp. 169–191.

75. Commoner, *The Closing Circle*, p. 299.

15

Ritual Is Essential

Dolores LaChapelle

Most native societies around the world had three common characteristics: they had an intimate, conscious relationship with their place; they were stable "sustainable" cultures, often lasting for thousands of years; and they had a rich ceremonial and ritual life. They saw these three as intimately connected. Out of the hundreds of examples of this, consider the following:

1. The Tukano Indians of the Northwest Amazon River basin, guided by their shamans who are conscious ecologists, make use of various myths and rituals that prevent over-hunting or over-fishing. They view their universe as a circuit of energy in which the entire cosmos participates. The basic circuit of energy consists of "a limited quantity of procreative energy that flows continually between man and animals, between society and nature." Reichel-Dolmatoff, the Colombian anthropologist, notes that the Tukano have very little interest in exploiting natural resources more effectively but are greatly interested in "accumulating more factual knowledge about biological reality and, above all, about knowing what the physical world requires from men."

2. The Kung people of the Kalahari desert have been living in exactly the same place for 11,000 years! They have very few material belongings but their ritual life is one of the most sophisticated of any group.

3. Roy Rappaport has shown that the rituals of the Tsembaga of

New Guinea allocate scarce protein for the humans who need it without causing irreversible damage to the land.

4. The longest inhabited place in the United States is the Hopi village of Oraibi. At certain times of the year they may spend up to half their time in ritual activity.

5. Upon the death of their old *cacique*, Santa Ana Pueblo in New Mexico recently elected a young man to take over as the new *cacique*. For the rest of his life he will do nothing else whatsoever but take care of the ritual life of the Pueblo. *All* his personal needs will be taken care of by the tribe. But he cannot travel any further than sixty miles or one hour distance. The distance has grown further with the use of cars but the time remains the same—one hour away from the Pueblo—his presence is that important to the ongoing life of the Pueblo. They know that it is ritual which embodies the people.

Our Western European industrial culture provides a striking contrast to all these examples. We have idolized ideals, rationality and a limited kind of "practicality," and have regarded the conscious rituals of these other cultures as at best frivolous curiosities. The results are all too evident. We've only been here a few hundred years and already we have done irreparable damage to vast areas of this country now called the U.S. As Gregory Bateson notes, "mere purposive rationality is necessarily pathogenic and destructive of life."

We have tried to relate to the world around us through only the left side of our brain, and we are clearly failing. If we are to re-establish a viable relationship, we will need to rediscover the wisdom of these other cultures who knew that their relationship to the land and to the natural world required the whole of their being. What we call their "ritual and ceremony" was a sophisticated social and spiritual technology, refined through many thousands of years of experience, that maintained their relationship much more successfully than we are.

The industrial humans have forgotten so much in the last 200 years that we hardly know where to begin. But it helps to begin remembering. In the first place, *all* traditional cultures, even our own long-ago Western European cultural ancestors, had seasonal festivals and rituals. The true origin of most of our modern major holidays dates back to these seasonal festivals. There are four major festivals: winter and summer solstice (when the sun reverses its travels) and spring and autumn equinox (when night and day are equal). But in between each of these major holidays are the "cross quarter days." For example, spring equinox comes around March 21 or 22 but spring is only barely beginning at that time in Europe. True spring— warm reliable spring—doesn't come until later. This is the cross quarter day—May 1—which Europe celebrated with maypoles, gathering flowers, and fertility rites. May became the month of Mary after the Christian church took over, and May crownings and processions were devoted to Mary instead of the old "earth goddesses." Summer solstice comes on June 21. The next cross quarter day is Lammas Day in early August. This is the only festival that our country does not celebrate in any way. The Church put the Feast of the Assumption on this day to honor Mary. Fall equinox comes on September 21—the cross quarter day is Hallowe'en, the ancient Samhain of the Celts. Then comes winter solstice—the sun's turn-around point from darkness to light. The cross quarter day between the solstice and spring equinox is in early February—now celebrated in the church as Candlemas.

The purpose of seasonal festivals is periodically to revive the *topocosm*. Gaster coined this word from the Greek—*topo* for place and *cosmos* for world order. *Topocosm* means "the world order of a particular place." The *topocosm* is the entire complex of any given locality conceived as a living organism—not just the human community but the total community—the plants, animals and soils of the place. The *topocosm* is not only the actual and present living community, but also that continuous entity of which the present community is but the current manifestation.

Seasonal festivals make use of myths, art, dance and games. All of these aspects of ritual serve to connect—to keep open the essential connections within ourselves. Festivals connect the conscious with the unconscious, the right and left hemispheres of the brain, the cortex with the older three brains (this includes the Oriental *tan tien* four fingers below the navel), as well as connecting the human with the non-human—the earth, the sky, the animals and plants.

The next step after seasonal rituals is to acknowledge the non-human co-inhabitants of your place. You can begin by looking into the records of the tribes of Indians who lived there and see what their totem was. Look into the accounts of the early explorers and very early settlers. Barry Lopez relates that the Eskimo told him that their totem animal was always the one who could teach them something they needed to learn.

Beginning in the Northwest, it is fitting that we talk of Salmon. Salmon is the totem animal for the North Pacific Rim. "Only Salmon, as a species, informs us humans, as a species, of the vastness and unity of the North Pacific Ocean and its rim. . . . Totemism is a method of perceiving power, goodness and mutuality in locale through the recognition of and respect for the vitality, spirit and interdependence of other species," as Linn House explains. For at least 20,000 years the Yurok, Chinook, Salish, Kwakiutl, Haida and Aleut on this side of the rim, and on the other rim of the Pacific, the Ainu (the primitives of Japan), ordered their daily lives according to the timing of the Salmon population.

Several years ago I did some in-depth study of Celtic myth and discovered that Salmon was the totem animal for the Celts, too. According to their myth, there was a sacred well situated under the sea where the sacred Salmon acquired their supernatural wisdom. The famous Celtic hero, Finn, traditionally obtained his wisdom when he sucked on the thumb he had just burnt when picking up the Salmon he cooked. It is not surprising that Salmon links all these areas. The North Pacific Rim and the British Isles are maritime climates in the northern half of the earth. Here is the perfect way to ritualize the link between planetary villagers around the earth—through their totem animal.

How can we learn from Salmon? One specific way is to reclaim our waterways so that Salmon can again flourish. If we reclaim the water so that Salmon can flourish we have reclaimed the soil, the plants and the other species of the ecosystem—restored them to aboriginal health. In so doing we would be restoring full health to our children as well.

Linn House feels that the people who live in or near the spawning ground of Salmon should form associations, not as law enforcement agencies such as the State Fish and Game Department, but as educational groups and providers of ritual and ceremony which would celebrate the interdependence of species. Linn was a Salmon fisherman on Guemes Island—he now lives in Northern California where he is restocking Salmon rivers.

What relevance does this kind of ritual have for people who live in the city? All of us need seasonal and nature rituals wherever we live, but let me give you a specifically urban example. Siena, Italy, with a population of about 59,000, has the lowest crime rate of any Western city of a comparable size. Delinquency, drug addiction and violence are virtually unknown. Class is not pitted against class nor young against old.

Why? Because it is a tribal, ritualized city organized around the *contrada* (clans)—with names such as Chiocciola, the Snail; Tartule, the Turtle, etc.—and the *Palio* (the annual horse race). The *contrada* function as independent city states. Each has its own flag, its own territorial boundaries, its own discrete identity, church songs, patron saint and rituals. Particular topographical features of each *contrada's* area are ritualized and mythologized. The ritualized city customs extend clear back to the worship of Diana, the Roman goddess of the moon. Her attributes were taken over by the worship of Mary when Christianity came in.

Many famous writers such as Henry James, Ezra Pound and Aldous Huxley sensed the energy of the city and its events and tried to write about it, but none of them even faintly grasped the year-long ritualized life behind it. About one week before the day of the *Palio* race, workmen from the city of Siena begin to bring yellow earth (*la terra* from the fields outside Siena) and spread it over the great central square, the Campo, thus linking the city with its origins in the earth of its *place*. In fact, any time during the course of the year when someone needs to be cheered up, the sad person is told not to worry because there will be "la terra in piazza" (soon there will be earth in the square).

The horse race serves two main purposes. In the intense rivalry surrounding the race, each *contrada* "rekindles its own sense of identity." The *Palio* also provides the Sienese with an outlet for their aggression and as such is a ritual war. The horse race grew out of games which were actually mimic battles and were used to mark the ends of religious festivals.

The *Palio* is truly a religious event. On this one day of the year the *contrada's* horse is brought into the church of its patron saint. In the act of blessing the horse, the *contrada* itself is blessed. This horse race is the community's greatest rite. "In the Palio, all the flames of Hell are transformed into the lights of Paradise," according to a local priest, Don Vittorio.

If we want to build a sustainable culture, it is not enough to "go back to the land." That's exactly where our pioneering ancestors lived and, as the famous Western painter Charles Russell said, "A pioneer is a man who comes to virgin country, traps off all the fur, kills off the wild meat, plows the roots up.... A pioneer destroys things and calls it civilization."

If we are to truly re-connect with the land, we need to change our perceptions and approach more than our location. As long as we limit ourselves to rationality and its limited sense of "practical-ity," we will be disconnected from the "deep ecology" of our place. As Heidegger explains: "Dwelling is not primarily inhabiting but taking care of and creating that space within which something comes into its own and flourishes." It takes both time and ritual for real dwelling. Likewise, as Roy Rappaport observes, "Knowledge will never replace respect in man's dealings with ecological systems, for the ecological systems in which man participates are likely to be so complex that he may never have sufficient comprehension of their content and structure to permit him to predict the outcome of many of his own acts."

Ritual is the focused way in which we both experience and express that respect. Ritual is essential because it is truly the pattern that connects. It provides communication at all levels—communication among all the systems within the individual human organism; between people within groups; between one group and another in a city; and throughout all these levels between the human and the non-human in the natural environment. Ritual provides us with a tool for learning to think logically, analogically and ecologically as we move toward a sustainable culture. Most important of all, perhaps, during rituals we have the experience, unique in our culture, of neither *opposing* nature nor *trying* to be in communion with nature, but of *finding* ourselves within nature, and that is the key to sustainable culture.

16

The Council of All Beings

Pat Fleming and Joanna Macy

Some twenty-five of us are gathered at a riverside wilderness site in
New South Wales, Australia. Last night we shared stories from our
experiences which awakened our concern—even anguish—over
what is happening to the natural world in our time. Although we
come from different backgrounds, we have this concern in common
and it has brought us here to work together. We want to strengthen
our courage and commitment to take action to heal our world.

We took time to honor that intention as we first sat down together
last night. It has called us to experiment with new ways of healing
our separation from nature which is at the root of the destruction
of the forests, the poisoning of the seas and soil. These ways of recon-
necting are not so new, after all, as Sheila, one of the weekend's co-
leaders, pointed out.

> As we explore different group processes for reconnecting with
> nature, we discover that it is not hard to find ones which work
> and feel authentic to us. This is not surprising when we consider
> the thousands of generations of humans who have participated
> in such processes, and how few the generations since we tem-
> porarily forgot them.

This morning we engaged in a number of group exercises to help
make us more conscious of our embeddedness in the web of life.
They helped us *remember* our bio-ecological history, as our species

and its forebears evolved through four and a half billion years of this planet's life. They helped us relax into our bodies, into our intuitive knowings, and our trust in each other. Now, after clearing up after lunch, we assemble to prepare ourselves for the promised ritual of the Council of All Beings.

Instead of beginning right away, we receive from our co-leader Frank an invitation to disperse and go off alone for an hour.

Find a place that feels special to you and simply be there, still and waiting. Let another life-form occur to you, one for whom you will speak at this afternoon's Council of All Beings. No need to try to make it happen. Just relax and let yourself be chosen by the life-form that wishes to speak through you. It could be a form of plant or animal life, or an ecological feature like a piece of land or a body of water. Often the first that occurs to you is what is right for you at this gathering.

Even before I sit down quietly on the warm sand beside the river I have a sense of the "being" that awaits to emerge in me. It is Mountain.

I relax and breathe in, breathe in Mountain. . . . I feel my rock-roots go deep deep down to where the Earth herself is very hot. My base is wide, very wide and solid. Storms come and go over my surface, leaving but a ruffle in my tree-skin. Even the occasional quake of Earth only causes me to shiver and sense more keenly my vitality. I am ancient. I am aware of the centuries, the millennia passing, many seasons and cycles of change. I feel very strong, able to withstand a great deal. Out of this strength I offer protection and shelter to many other life-forms. I also offer inspiration and challenge: the call to come and know me better, to explore my rocks, valleys, and tree-lined rivers. I feel great peace within—immovable, beyond time. I offer this to all who rest upon me. Yet humans, like ants upon my flanks, can be so intrusive, abusive. They gouge my bones, strip flesh, bring danger to my inhabitants. I must speak of this at the Council this afternoon.

Brmm, brmm . . . the drum calls us back. It summons us all back to take time to make our masks and to explore further together our various life-forms. In companionable silence our hands reach for paper, colors, paste-pot. Under the rustling trees are sounds of cutting, folding, breathing.

Brmm, brmm . . . the drum calls us once again, this time to enter the ritual ground and convene as the Council of All Beings. Wearing my Mountain mask of earth, stone, leaves and grasses, I move heavily and slowly toward the ritual ground. The water of the stream that borders the site is cool, washing away the old, preparing us for the new, the unknown. As we gather in a large circle I look around at all the assembled Beings—such an array of forms and colors, some brash, some shy and subtle. An air of uncertainty, yet expectation hangs between us. The gum trees and old fella-grass tree bounding the site rustle in anticipation. A white egret flies past us upstream, flapping its well wishes to our gathering. I settle into the sand, solid, strong, waiting—ready for eternity if need be.

Frank briefs us first on the structure of the ritual. I recognize a blending of different native traditions of our planet's peoples. Through fire and water, we will ritually cleanse ourselves and the ritual ground. To acknowledge the full breadth of our concerns, we will invoke and invite into our circle the earth powers and beings that surround our lives in this space and time: the powers of the four directions and the beings of the three times. Then as the Council proper begins, we will, as the life-forms we have assumed, speak spontaneously, letting be said what needs to be said.

These utterances, we are told, will fall into three stages. From the perspective of the other life-forms we will speak spontaneously among ourselves. We will say why we have come to the Council and be free to express our confusion, our grief and anger and fear. Then, after a while, to the signal of a drumbeat, five or six of us at a time will move to sit in the center of the circle to listen in silence as humans. We'll each have the opportunity to shift between human and non-human roles. And lastly we will have the chance to offer to the humans (and receive as humans) the powers that are needed to stop the destruction of our world. Frank offers this overview of the structure to help us feel at ease with the structure of the ritual. He adds that fortunately we do not need to hold its sequence in our minds, for we will respond naturally as the ritual unfolds.

Now with the slow beat of drum and the lighting of fragrant leaves, the Council ritual begins. An abalone shell with burning sage and cedar passes from hand to hand; we inhale the sweet pungent smoke, waving it over our faces. We acknowledge our kinship with fire. Next a glass bowl of fresh water comes round the circle. Each dips into it to anoint the head of the next person, acknowledging our need for cleansing and refreshment.

As the four directions are invoked, we turn to the East, to the South, to the West and North. Drawing on the ancient lore of the Medicine Wheel, we all face in each direction, arms upraised, as one of our number evokes in turn and aloud the meanings it can hold. "We invoke and invite the power of the East ... the power of the rising sun, of new beginnings, the farsightedness of eagle...." As Mountain I feel special kinship with the North, "powers of stillness and introspection, of waiting and endurance...."

After each invocation, we all join in with a simple, deep, two-tone refrain: "Gather with us now in this hour; join with us now in this place."

Frank, as the ritual leader, now helps us invite the Beings of the Three Times.

> We invite into our circle all those who have gone before. All you who have lived upon Earth and loved and nurtured it, we remember you and call on your wisdoms and your hopes. You, our ancestors and teachers, we ask you to be with us in this time. We seek strength and counsel for the saving of the Earth you loved. We say your names.

Spontaneously, and all intermingled from around the circle, comes the murmur of randomly spoken names—names of Jesus and the Buddha, of Martin Luther King and grandparents, school teachers and spiritual guides.

> We invoke also the beings of the present time. We invite into this circle our families and friends, our neighbors and co-workers and those who have made it possible for us to be here at this camp. You share our hope and fears, and we are here for you, too.

And from the circle rises again a murmur of names ... "Peter, Adele, Susan, George Hawke, Mikhail Gorbachev."

> Lastly we invoke you beings of the future time, you who are waiting to be born. We cannot say your names, for you have none yet that we can know. Yet it is for you that we work to preserve this beautiful planet. And it is from you, too, that we need help to do what must be done.

After a moment of silence, a silence for the generations we hope will come after us, we chant again the refrain that has followed each invocation.

"Gather with us now in this hour; join with us now in this place."

Now we are ready to speak as a Council. We sit and take our masks. We ease out of our solely human identification; we settle into the life-forms that have come to us and that seek expression.

We hold a roll-call of the assembled beings. One by one around the circle, speaking through our masks, we identify ourselves: "I am wolf and I speak for the wolf people." "I am wild goose and I speak for all migratory birds." "I am wheat and I speak for all cultivated grains."

"We meet," Frank says, having donned his mask of prickly stalks and leaves,

> because our planet is in trouble. We meet to say what is happening to ourselves and our world. I come to this Council as weeds. Weeds, a name humans give to plants they do not use. I am vigorous, strong. I love to thrust and push and seed—even through concrete. Pushing through paving I bring moisture and life. I heal the burned and wounded earth. Yet I am doused with poison now and crushed, as are creatures who live in and through me.

In acknowledgment we all reply, "We hear you, Weeds."

> I am woompoo pigeon. Coooo. I live in one of the last pockets of rainforest. I call my song softly through the giant trees and the cool green light. Yet I no longer get a reply. Where are my kind? Where have they gone? I hear only the echo of my own call. I am frightened; that's why I'm here.

"We hear you woompoo pigeon."

> I am black and white cow, fenced in a paddock, far from grass, standing in my own shit. My calves are taken from me, and instead cold metal machines are clamped to my teats. I call and call, but my young never return. Where do they go? What happens to them?

"We hear you, cow."

> The shells of my eggs are so thin and brittle now, they break before my young are ready to hatch. I fear there is poison in my very bones.

"We hear you, wild goose."
One by one they speak and are heard. Rainforest. Wombat. Dead leaf. Condor. Mud. Wild flower.
A soft voice says,

> They call me slug—that doesn't bother me. I just slip along real slow and gentle, nubbling the leaves as I go. But, d'you know, just for this, I get mangled and chopped up, squashed without even a "How's your uncle?" What have I done to deserve this?

Laughter and sympathy; "We hear you, slug."
Red kangaroo, lichen, wild pig, bottle-nosed dolphin.
"We hear you, we hear you."
I know that it is my time to speak out.

> I am Mountain. I am ancient and strong and solid, built to endure. But now I am being dynamited and mined, my forest skin is being torn off me, my top-soil washed away, my streams and rivers choked. I've a great deal to address to the humans today.

"We hear you, Mountain."
The drum beats again, announcing the next stage of the Council. It summons humans to enter the circle to listen. Five or six of the beings put aside their masks and move to the center. Sitting back to back facing outward, they attend silently as the Council continues. When the drum beats again after several more beings have spoken, they return to the periphery to be replaced by others; and the

process continues until each of us will have had the chance to listen as a human.

"Hear us, humans," says Weeds.

> This is our world, too. And we've been here a lot longer than you. For millions of years we've been raising our young, rich in our ways and wisdom. Yet now our days are numbered because of what you are doing. Be still for once, and listen to us.

> I am Rainforest. Counted in your human years I am over a hundred thirty million years old. If I were one of your buildings, you would take precious care of me. But instead you destroy me. For newsprint and cheap hamburgers you lay me waste. You destroy me so carelessly, tearing down so many of my trees for a few planks, leaving the rest to rot or burn. You push needless roads through me, followed by empty-hearted real-estate grabbers who purport to own me. You cause my thick layer of precious topsoil to wash away, destroying the coral reefs that fringe me. I can't stand your screaming machines which tear through my trunks, rip my flesh, reducing hundreds of years of slow growth to sawdust and furniture. How dare you!

Standing up, majestic in his anger, Rainforest continues.

> Your greed and folly shortens your own life as a species. When you leave me wasted and smoldering, you foretell your own death. Don't you know that it is from me that you have come? Without my green world your spirit will shrivel, without the oxygen my plant life exhales, you'll have nothing to breathe. You need me as much as your own lungs. I *am* your lungs.

> Oh, humans, as Clean Water I was a bearer of life and nourishment. Look at what I bear now that you've poured your wastes and poisons into me. I am ashamed and want to stop flowing, for I have become a carrier of sickness and death.

Possum holds up her hand.

> See my hand, humans? It resembles yours. From its print on the soft soil you can tell where I have passed. What mark on this earth will you leave behind you?

Brmm, brmm ... the drum sounds into the circle. The humans in the center, looking relieved to be leaving, return to the wider circle and resume the masks of their other life-form. A half-dozen others move to the middle as humans and sit close together, some holding each others' hands as they listen.

> I am Bottle-nosed Dolphin. I love to roll and leap and play. Yes, humans, to play with you, too, when I can trust you, for we feel great affinity with you. But in the gill-nets you use we tangle and drown. Taking cruel advantage of our friendliness, you use us for military experiments, fix monitors and transmitters on our backs.
>
> You wall us into your sea parks for show. You deny us the chance to swim free and with our kind. I speak for all captive beings! Find your own freedom in honoring ours.
>
> I feel myself beginning to boil inside and know again I must speak.

> Humans! I, Mountain, am speaking. You cannot ignore me! I have been with you since your very beginnings and long before. For millennia your ancestors venerated my holy places, found wisdom in my heights. I gave you shelter and far vision. Now, in return, you ravage me. You dig and gouge for the jewel in the stone, for the ore in my veins. Stripping my forests, you take away my capacity to hold water and to release it slowly. See the silted rivers? See the floods? Can't you see? In destroying me you destroy yourselves. For Gaia's sake, wake up!

> Humans, I am Lichen. Slowly over the ages I turn rock into soil. I thought nothing could ever stop me. Till now. Now I sicken with your acid rain.

> Look on me, humans. I am the last wild Condor of that part of the Earth you call California. I was captured a few days ago—"for your own good," you tell me. Look long and hard at me, at the stretch of my wings, at the glisten of my feathers, the gleaming of my eye. Look now, for I shall not be here for your children to see.

"We hear you Dolphin, we hear you Rainforest ... we hear you Mountain, Lichen, Condor...."

One by one our stories pour out, filled with pain, with anger, and, occasionally, with humor. We all report how rapidly and radically humans are affecting our lives and our chances for survival. Yet the words carry, too, a sense of kindred spirit, for we are all of the same Earth.

When the flow of expression begins to subside, our ritual leader Frank, taking off his mask as Weeds, comes into the center of the circle. It is the first time in the ritual that we hear someone speak as a human.

> We hear you, fellow beings. It has been painful to hear, but we thank you for your honesty. We *see* what we're destroying, we're in trouble and we're scared. What we've let loose upon the world has such momentum, we feel overwhelmed. Don't leave us alone— we need your help, and for your own survival too. Are there powers and strengths you can share with us in this hard time?

No other signal or instruction is needed to shift the mood of the Council. The grim reports and chastisements give way to a spontaneous sharing of gifts.

> As Slug, I go through life slowly, keeping close to the ground. I offer you just that, humans. You go too far, too fast for anyone's good. Know carefully and closely the ground you travel on.

Water says,

> I flow on and on. I deal with obstacles by persistence and flexibility. Take those two gifts for your lives and your work for the planet.

> I, Condor, give you my keen, far-seeing eye. I see at a great distance what is there and what is coming. Use that power to look ahead beyond your day's busy-ness, to heed what you see and plan.

"Thank you, Slug ... thank you. Water ... Condor," murmur the humans.

One after another the beings offer their particular powers to the humans in the center. After speaking, each leaves its mask in the

outer circle and joins the humans in the center, to receive empowerment as a human from the other life-forms.

> As Lichen, I work with time, great stretches of time. I know time
> is my friend. I give you that patience for the long haul. I would
> relieve you of haste.

> I, Rainforest, offer you my powers in creating balance and har-
> mony that enable many life-forms to live together. Out of this
> balance and symbiosis new, diverse life can spring. This I can offer
> to you.

> As Dead Leaf, I would free you from your fear of death. My drop-
> ping, crumbling, molding allows fresh growth. Maybe if you were
> less afraid of death, you'd be readier to live. I offer you com-
> panionship with death, as you work with the natural, healing
> cycles of life.

Wildflower speaks.

> I offer my fragrance and sweet face to call you back to life's beauty.
> Take time to notice me and I'll let you fall in love again with life.
> This is my gift.

I feel Mountain wanting to speak through me again.

> Humans, I offer you my deep peace. Come to me at any time to
> rest, to dream. Without dreams you may lose your vision and
> your hope. Come, too, for my strength and steadfastness, when-
> ever you need them.

I take off my mask and join the group of fellow-humans in the center. Hands reach out to pull me in close. I feel how warm and wel-
coming is the touch of human skin. I am beginning to gain a fresh
recognition of our strengths. For all the gifts that the beings offer
are already within us as potentialities, otherwise we would not have
been able to articulate them.

The last of the beings gives its blessings. Frank has taken up his
mask again and speaks:

> I offer you our power as Weeds—that of tenacity. However hard
> the ground, we don't give up! We know how to keep at it, slowly

at first, resting when needed, keeping on—until suddenly—crack! and we're in the sunlight again. We keep on growing wherever we are. This is what we share with you—our persistence.

We thank him and pull him into our midst. A wordless sounding arises. Grasping hands, we stand and begin to move outward in a circle, laughing and humming. Sheila leads the long line of us back in upon itself, coiling gradually tighter and closer around ourselves into a group embrace. It is the ancient form from this land's aboriginal tradition known as a "humming bee." The close intertwined embrace, cheeks against shoulders, skin on skin, feels good, as the humming of our throats and chests vibrates through us. It is as if we are one organism.

The hum turns to singing. Someone takes the drum. To its rhythm some move and dance, leaping and swaying and stamping the ground. Others move off among the trees and down to the waterside, to be quiet with themselves and what has happened.

Later, as the sun sets, we reassemble to release the life-forms that we have allowed to speak through us. A fire has been kindled in the growing darkness. One by one we come forward with our masks and put them in the flames, honoring the beings they symbolized and letting them burn. "Thank you, Condor." "Thank you, Mountain."

Tomorrow the circle will meet again, to speak of the changes we as humans will work for in our lives and in our world. Then we will make plans for action, hatch strategies, concoct ways of supporting each other. Right now it is good simply to rest upon our ground and watch the masks curl and crackle as they catch fire.

17

Ecology, Place, and the Awakening of Compassion

Gary Snyder

Oecology, as it used to be spelled, is a scientific study of relationships, energy-transfers, mutualities, connections, and cause-and-effect networks within natural systems. By virtue of its finding, it has become a discipline that informs the world about the danger of the breakdown of the biological world. In a way, it is to Euro-American global economic development as anthropology used to be to colonialism. That is to say, a kind of counter-science generated by the abuses of the development culture (and capable of being misused by unscrupulous science mercenaries in the service of the development culture). The word "ecological" has also come to be used to mean something like "environmentally conscious."

The scientist, we are told, seeks to be objective. Objectivity is a semi-subjective affair, and although one would aspire to see with the distant and detached eye of a pure observer, when looking at natural systems the observer is not only affecting the system, he or she is inevitably *part* of it. The biological world with its ecological interactions is *this* world, our very own world. Thus, ecology (with its root meaning of "household science") is very close to economics, with its root meaning of "household management." Human beings, biology and ecology tell us, are located completely within the sphere of nature. Social organization, language, cultural practices, and other features that we take to be distinguishing characteristics of the human species are also within the larger sphere of nature.

To thus locate the human species as being so completely within "nature" is an upsetting step in terms of the long traditions of Euro-American thought. Darwin proposed evolutionary and genetic kinship with other species. This is an idea that has been accepted intellectually, but not personally and emotionally, by most people. Social Darwinism flourished for a while as a popular ideology justifying nineteenth-century imperialism and capitalism, with an admiring emphasis on competition. The science of ecology corrects that emphasis and goes a step further. It acknowledges the competitive side of the process, but also brings forward the co-evolutionary, cooperative side of interactions in living systems. Ecological science shows us that nature is not just an assembly of separate species all competing with each other for survival (an urban interpretation of the world?)—but that the organic world is made up of many communities of diverse beings in which the species all play different but essential roles. It could be seen as a village model of the world.

An ecosystem is a kind of mandala in which there are multiple relations that are all powerful and instructive. Each figure in the mandala—a little mouse or bird (or little god or demon figure) has an important position and a role to play. Although an ecosystem can be described as hierarchical in terms of energy-flow, from the standpoint of the whole all of its members are equal.

But we must not sentimentalize this. A key transaction in natural systems is energy-exchange, which means the food-chains and the food-webs, which means that many living beings live by eating other beings. Our bodies—or the energy they represent—are thus continually being passed around. We are all guests at the feast, and we are also the meal! All of biological nature can be seen as an enormous *puja*, a ceremony of offering and sharing.

The intimate perception of interconnection, frailty, inevitable impermanence and pain (and the continuity of grand process and its ultimate emptiness) is an experience that awakens the heart of compassion. It is the insight of *bodhicitta* that Shantideva wrote of so eloquently. It is the simultaneous awakening of a personal aspiration for enlightenment and a profound concern for others.

Ecological science clearly throws considerable light on the fundamental questions of who we are, how we exist, and where we belong. It suggests a leap into a larger sense of self and family. It seems clear enough that a consequence of our human interdependence should be a social ethic of mutual respect, and a commitment to solving conflict as peacefully as possible. As we know, history tells a different story. Nonetheless, we must forge on to ask the next question: How do we encourage and develop an ethic that goes beyond intra-human obligations and includes non-human nature? The last 200 years of scientific and social materialism, with some exceptions, has declared our universe to be without soul and without value except as given value by human activities. The ideology of development is solidly founded on this assumption. Although there is a tentative effort among Christians and Jews of good will to enlarge their sense of ethics to include nature (and there have been a few conferences on "eco-Christianity"), the mainstream of Euro-American spirituality is decidedly human-centered.

Asian thought-systems (although not ideal) serve the natural world a little better. Chinese Daoism, the Sanatana ("eternal") Dharma of India, and the Buddhadharma of much of the rest of Asia all see humanity as part of nature. All living creatures are equal actors in the divine drama of Awakening. As Tashi Rapges said, the spontaneous awakening of compassion for others instantly starts one on the path toward enlightenment. They are not two. In our contemporary world, an ethic of concern for the non-human arrives not a moment too soon. The biological health of the planet is in trouble. Many larger animals are in danger of becoming extinct, and whole ecosystems with their *lakhs* of little living creatures are being eliminated. Scientific ecology, in witness to this, has brought forth the crisis-discipline of Conservation Biology, with its focus on preserving biodiversity. Biodiversity issues now bring local people, industries, and governments into direct and passionate dialogue over issues involving fisheries, marine mammals, large rare vertebrates, obscure species of owls, the building of huge dams or road systems—as never before.

The awakening of the Mind of Compassion is a universally known human experience, and is not created by Buddhism or any other particular tradition. It is an immediate experience of great impact, and Christians, Jews, Muslims, Communists, and Capitalists will often arrive at it directly—in spite of the silence of their own religions or teaching on such matters. The experience may often be completely without obvious ethical content, a moment of leaving hard ego self behind while *just seeing*, just being, at one with some other.

Much of India and the Far East subscribes in theory at least to the basic precept of non-harming. *Ahimsa*, nonviolence, harmlessness, is described as meaning "Cause the least possible harm in every situation." Even as we acknowledge the basic truth that every one of us lives by causing some harm, we can consciously amend our behavior to practically reduce the amount of damage we might do, without being drawn into needless feelings of guilt.

Keeping nature and culture healthy in this complicated world calls us to a kind of political and social activism. We must study the ways to influence public policy. In the western hemisphere we have some large and well-organized national and international environmental organizations. They do needed work, but are inevitably living close to the centers of power, where they lobby politicians and negotiate with corporations. In consequence, they do not always understand and sympathize with the situations of local people, village economics, tribal territories, or impoverished wage-workers. Many scientists and environmental workers lose track of that heart of compassion, and their memory of wild nature.

The actualization of the spiritual and political implications of ecology—that it be more than rhetoric or ideas—must take place, place by place. Nature happens, culture happens, *somewhere*. This grounding is the source of bioregional community politics. Joanna Macy and John Seed have worked with the image of a "Council of All Beings." The idea of a *Village* Council of All Beings suggests that we can get specific. Think of a village that includes the trees and birds, the sheep, goats, cows, and yaks and the wild animals of the high pastures (ibex, argali, antelope, wild yak) as members of the community. And whose councils, in some sense, give them voice.

Then to provide space. Any of these Ladakhi and Tibetan village territories certainly should include the distant communal pastures (*p'u*) and the sub-watersheds as well as the cultivated fields and households. In the case of Ladakh, and indeed all of India, when a village is dealing with government or corporation representatives it should insist that the "locally used territory" embraces the whole local watershed. Otherwise, as we have too often seen, the government agencies or business forces manage to co-opt the local hinterland as private or "national" property, and relentlessly develop it according to an industrial model.

We need an education for the young people that gives them pride in their culture and their place, even while showing them the way into modern information pathways and introducing them to the complicated dynamics of world markets. They must become well informed about the workings of governments, banking, and economics, those despised, but essential mysteries. We need an education that places them firmly within biology, but also gives a picture of human cultural affairs and accomplishments over the millennia. (There is scarcely a tribal or village culture that doesn't have some sort of music, drama, craft, and story that it can be proud of when measured against the rest of the world.) We must further a spiritual education that helps children appreciate the full interconnectedness of life and encourages a biologically informed ethic of non-harming.

All of us can be as placed and grounded as a willow tree along the streams—and also as free and fluid in the life of the whole planet as the water in the water cycle that passes through all forms and positions roughly every two million years. Our finite bodies and inevitable membership in cultures and regions must be taken as a valuable and positive condition of existence. Mind is fluid, nature is porous, and both biologically and culturally we are always fully part of the whole. This ancient nation of Ladakh has always had such people living in it. Some of the beautiful young men and women here today will master the modern world, keep up the Dharma and continue to be true people of Ladakh.

18

Four Forms of Ecological Consciousness Reconsidered

John Rodman

The primary purpose of this essay is to describe and evaluate as succinctly as possible four currents of thought discernible in the history
of the contemporary environmental movement. The secondary purpose is to recommend the still-emergent fourth form (Ecological Sensibility) as the starting point for a general environmental ethic. Along
the way, I hope to suggest something of the complexity and ambiguity of the various forms—qualities sometimes lost in the rush to
condemn the "shallow" and extoll the "deep."

1. Resource Conservation

The basic thrust of the Resource Conservation standpoint, taken in
its turn-of-the-century context and seen as its advocates saw it, was
to restrain the reckless exploitation of forests, wildlife, soils, etc.,
characteristic of the pioneer stage of modern social development by
imposing ethical and legal requirements that "natural resources" be
used "wisely," meaning (in Gifford Pinchot's words) that they should
be used "for the greatest good of the greatest number" (of humans),
as distinct from being used to profit a few, and that the good should
be considered over "the long run," that is, in terms of a sustainable
society. Now that the novelty of this standpoint has worn off and
more radical views have arisen, it is clear that the ethic of "wise use"
remained within the worldview of anthropocentric utilitarianism,
since it assumed (without arguing the point) that nonhuman natural

entities and systems had only instrumental value as (actual or potential) "resources" for human use, so that the only reasons for humans to restrain their treatment of nonhuman nature were prudential ones flowing from considerations of enlightened self-interest.

What is a self and what is an interest, however, are not exactly given once and for all. To Pinchot's identification of human interests with economic prosperity and national power others have subsequently added such things as aesthetic enjoyment, scientific knowledge, and (more recently) biological survival. To Pinchot's and Theodore Roosevelt's extension of the self in space and time to comprise a national society and to include the interests of overlapping generations (ourselves, "our children, and our children's children"), others have superadded the notion of obligations to a remoter posterity to which we are linked by the half-life of radioactive nuclear waste. Clearly, it is possible to engage in a good deal of persuasion by using and extending key terms and commitments within the Resource Conservation standpoint, as, for example, when Aldo Leopold redefined wealth and poverty in aesthetic terms so that an economically poor landscape might be seen as rich in beauty and therefore worth preserving. In the case of aesthetic contemplation and certain kinds of disinterested scientific knowledge (e.g., that of the field naturalist), we approach the boundary of the Resource Conservation position in the sense that these are undeniable human interests that are so distant from the original and core (economic) sense of "use" (which involved the damming, logging, bulldozing, and transformation of nature into manufactured products) that the non- (or at least significantly less-) exploitative, more respectful senses of "use" can provide bridges for crossing over into the notion that there is intrinsic value in (some) natural entities and systems, which, after all, are beautiful or interesting to us partly in virtue of qualities that inhere *in them.*

Insofar as the Resource Conservation standpoint is retained in its core assumptions, however, it is vulnerable on several related grounds. First, the reduction of intrinsic value to human beings and the satisfaction of their interests is arbitrary, since it is neither necessary (for there are other human cultures that have not so reduced value), nor justified (for nobody has yet successfully identified an observable, morally relevant quality that both includes all humans and excludes all nonhumans). Second, the commitment to maximizing value through maximizing human use leads logically and in practice to an unconstrained total-use approach, whose upshot is to leave nothing in its natural condition (for that would be a kind of "waste," and waste should be eliminated); all rivers should be dammed for irrigation and hydropower, and all native forests replaced with monocultural tree plantations managed for "harvest." Given the arbitrariness of the first principle, the second amounts to an unjustifiable species imperialism.

Granted all this, it would be unfortunate if the contemporary environmental movement turned out to be simply what Stewart Udall once called it, "the third wave of conservation." Yet it is also important to recognize the (limited) validity of the Resource Conservation standpoint in terms of its historical thrust. It emerged, in large part, as an attempt to constrain the destructive environmental impact of individuals and corporations who exploited nature for profit without sufficient regard for the larger social good or for the welfare of future generations. That issue has not died or become unimportant because broader ones have arisen. Some acts are wrong on several grounds: this is what makes possible the formation of honest coalitions, which are indispensable to political efficacy. Put in the most general way, the original thrust of the Resource Conservation movement was to enlarge in space and time the class of beings whose good ought to be taken into account by decision-makers, and to draw from that some conclusions about appropriate limits on human conduct. That Pinchot and his followers were inhibited by an unquestioning, almost unconscious, fidelity to an anthropocentric reduction of intrinsic value is cause for regret—and for criticism, but the direction of thrust warrants respect from even the most radical environmentalists. In retrospect, the Resource Conservation standpoint appears to have been an early ideological adaptation on the part of a society that was still in the pioneering or colonizing stage of succession but had begun to get glimpses of natural limits that would require different norms of conduct for the society to become sustainable at a steady-state level. How different those norms might have to be was not yet clear.

2. Wilderness Preservation

At approximately the same time (1890–1914) that the Conservation movement was defining itself against the forces of unbridled resource exploitation, the Wilderness Preservation tradition, represented in part by John Muir and the Sierra Club, was also emerging as a social force.[1] At first allied with Pinchot against the common enemy and under the common banner of "conservation," Muir parted ways with him over issues such as the leasing of lands in the federal forest reserves to commercial grazing corporations and over the proposal to dam Hetch Hetchy Valley to make a reservoir for the growing population of San Francisco. What seemed a wise use to Pinchot, weighing the number of city dwellers, seemed a "desecration" to Muir. In contrast to the essentially economic language of Resource Conservation, Preservationists tended to articulate their vision in predominantly religious and aesthetic terms. Beneath Muir's somewhat parochial tendency to depict particular landscapes as "temples" and "cathedrals" lay both a rather pantheistic view of nature as animated by a divine power ("God" being sometimes equated with "Beauty") and also a very ancient and widespread notion that certain natural areas were sacred places where human beings could encounter the holy. This has been a potent and enduring strand in American environmentalism, going back to Emerson and Thoreau. Sociologist Arthur St. George concluded his study of Sierra Club members in the 1970s with the finding that their basic values were (still) "religious and aesthetic." Surely Muir would have applauded, had he been alive in 1966, when the Sierra Club took out full-page newspaper ads satirizing the Bureau of Reclamation's claim that damming the Grand Canyon for irrigation, etc., would enhance its recreational value: "Should we also flood the Sistine Chapel so tourists can get nearer the ceiling?" The implied analogy between Grand Canyon and a chapel full of priceless works of art was not only clever but faithful to a long-standing tradition.

When the religious element in the Preservationist tradition becomes explicit, some philosophers are embarrassed and inclined to dismiss it in a sentence or two as "mysticism," "superstition," "irrational-ism," or at least as an unnecessary and misleading backdrop that disguises the real thing at stake and will surely lead other people to discount the case for preserving wilderness. Perhaps this reaction stems, at least in part, from assuming that the important issues are metaphysical ones (Does God exist? Is (S)he present in nature?) instead of experiential ones (What do people tend to feel in certain natural settings, and, given their cultural background, how do they tend to articulate those feelings?). I have given elsewhere[2] a rather psychological interpretation of the wilderness experience, which may help salvage it as an intelligible phenomenon for secular minds at the cost of prompting the question whether the value of Nature for the Preservationist lies solely in its therapeutic utility for what Muir described as "thousands of tired, nerve-shaken, over-civilized peo-ple." If the answer is yes, then Wilderness Preservation should be seen as a genteel variant on Resource Conservation. There is no doubt that the passage of time affects our perspective on this kind of thing. The more distant we are from Pinchot's struggle against "reckless" resource exploitation, the easier it is to see Resource Conservation as simply a more prudent form of resource exploita-tion. The more distant we are from the split between Muir and Pin-chot, and the more aware we have become of alternative positions that assert the intrinsic value of nature with less ambiguity, the more inclined we are to see this once world-historical schism as a family quarrel between advocates of two different forms of human use—economic and religio-aesthetic. On the other hand, if Muir had been asked outright whether Yosemite had value in itself, or for its own sake, independent of there being any actual or potential people to experience it, he would have surely said that it did. Whether this would be value independent of a divine, creative force lurking in the universe, however, is doubtful. The Preservationist notion of nature's value is thus in continual danger of being reduced by critics one way or another—either to subjective human experience or to an (allegedly) objective deity that is manifested in nature. For the person unsatis-fied by the former but unable to believe in the latter, the Preserva-tionist view poses serious difficulties.

Since this issue can be argued back and forth without any clear conclusion being reached, I should like to approach it from a slightly different tack, focusing on the aesthetic that mediates the experience of the holy. In general, aesthetic considerations are very subjective and therefore shaky foundations on which to base any kind of ethic. The specific problem with the Preservationist tradition emerges when we consider that until the eighteenth century most Europeans, including early American settlers, viewed "wilderness" and especially mountains with horror and dread, and that the emergence of wilderness-appreciation and mountain-climbing as cults accompanied the emergence of the Romantic aesthetic of "the sublime and the beautiful," especially the sublime, meaning essentially the feeling of awe experienced in the presence of overwhelming power, magnitude, or antiquity. For two centuries the aesthetic of the sublime has, in effect, defined outdoor sacred space for people of European background, who, consequently, have often had difficulty relating to the claims of aboriginal peoples whose sacred places do not always conform to our ideas of what a sacred place is like. Preservationist efforts have tended to focus on protecting more-or-less pristine wilderness at the expense of disturbed areas, and on saving the Sierras, the Grand Canyons, and the giant sequoias of life as the natural environments of peak experiences, while the marshes and the brushlands have gradually disappeared under the impact of less exalted forms of human "growth." Of course, it is not *a priori* false that some natural areas are more valuable than others; and, given environmentalists' limits of numbers, time, and energy, priorities may be a practical necessity. Still, the particular pattern of discrimination exhibited by the history of the Preservation movement stems all too clearly from historically transient aesthetic tastes and bears too little relation to sustaining healthy ecological systems.

An aesthetic of nature can evolve (e.g., to incorporate an appreciation of small organisms and of complex and intricate interrelatedness), and an organized movement can evolve away from aesthetics. As Sierra Club members become also Friends of the Earth, they lose a certain ferocity of commitment that is probably possible only to places that are special, while they gain a more generalized commitment to defending ecological values. In special cases, when grand, ancient, or otherwise awe-inspiring wild areas are threatened, the Preservationist tradition still provides a wellspring of powerful feeling and rhetoric that can be tapped. But, even within a more-or-less pantheistic framework, sacred places exist in contrast to profane ones, and the Preservationist tradition cannot—without serious attenuation of its core values or indulgence in gross eclecticism— do the work of a general environmental ethic.

3. Moral Extensionism

Moral Extensionism (which I called "Nature Moralism" in earlier papers) is an appropriately awkward term invented to designate a wide range of positions whose common characteristic is that they contend that humans have duties not only concerning but also directly to (some) nonhuman natural entities, that these duties derive from rights possessed by the natural entities, and that these rights are grounded in the possession by the natural entities of an intrinsically valuable quality such as intelligence, sentience, or consciousness. Quite different versions of this position can be found in the writings of such people as John Lilly, Peter Singer, Christopher Stone, and certain philosophers in the tradition of Whitehead (notably, Charles Hartshorne and John Cobb). All these writers *appear* to break with the anthropocentric bias of Resource Conservation and to resolve the ambiguity of Preservationism by clearly attributing intrinsic value to (at least some) natural items in their own right. The ground for human self-restraint towards nonhuman nature thus becomes moral in a strict sense (respect for rights) rather than prudential or reverential. Yet more radical environmentalists (e.g., the Routleys, Rodman, Callicott, Sessions, Devall et al.) object that the break with anthropocentrism and the resolution of the ambiguity are incomplete, and that all the variants of this position are open to the criticism that they merely "extend" (rather than seriously question or radically change) conventional anthropocentric ethics, so that they are vulnerable to revised versions of the central objection to the Resource Conservation standpoint, namely, that they are chauvinistic, imperialistic, etc.

Consider, for a starter, John Lilly's view that we ought to protect dolphins because they are very intelligent. Or consider, further along the spectrum of variants on this position, Peter Singer's argument that all sentient beings (animals down through shrimps) have an equal right to have their interests taken into consideration by humans who are making decisions that might cause them pain (pain being bad, and acts that cause unnecessary pain being wrong). Then consider, at the far end of the spectrum, the claim of various writers that all natural entities, including plants and rocks, have certain rights (e.g., a right to live and flourish) because they all possess some trait such as consciousness (though some possess it more fully than others). In the first two cases the scope of moral concern is extended to include, besides humans, certain classes of nonhumans that are like humans (with regard to the specified quality), while the vast bulk of nature is left in a condition of unredeemed thinghood. In Singer's version anthropocentrism has widened out to a kind of zoocentric sentientism, and we are asked to assume that the sole value of rainforest plant communities consists in being a natural resource for birds, possums, veneer manufacturers, and other sentient beings. In the third case, we are asked to adopt the implausible assumption that rocks (for example) are conscious. In all three cases we end up with an only slightly modified version of the conventional hierarchy of moral worth that locates humans at the top of the scale (of intelligence, consciousness, sentience), followed by "higher" animals, "lower" animals, plants, rocks and so forth. "Subhumans" may now be accorded rights, but we should not be surprised if their interests are normally overridden by the weightier interests of humans, for the choice of the quality to define the extended base class of those entitled to moral consideration has weighted the scales in that way.

Moreover, extensionist positions tend (when consistent, at least) to perpetuate the atomistic metaphysics that is so deeply imbedded in modern culture, locating intrinsic value only or primarily in individual persons, animals, plants, etc., rather than in communities or ecosystems, since individuals are our paradigmatic entities for thinking, being conscious, and feeling pain. Yet it seems bizarre to try to account wholly for the value of a forest or a swamp by itemizing and adding up the values of all the individual members. And it is not clear that rights and duties (of which our ideas are fundamentally individualistic) can be applied to ecosystem relationships without falling into absurdity. Pretty clearly what has happened is that, after both the prudential and the reverential stages of ideological adaptation represented by Resource Conservation and Wilderness Preservation came to seem inadequate, a more radical claim that nature had value "in its own right" seemed in order. Many of the attempts to make that claim plausible have, however, tried to extend the sphere of intrinsic value and therefore of obligatory moral concern by assimilating (parts of) nature to inappropriate models, without rethinking very thoroughly either the assumptions of conventional ethics or the ways in which we perceive and interpret the natural world. It is probably a safe maxim that there will be no revolution in ethics without a revolution in perception.

4. Ecological Sensibility

The last form that I shall discuss is still emergent, so that description is not easily separated from prescription. The term "sensibility" is chosen to suggest a complex pattern of perceptions, attitudes, and judgments which, if fully developed, would constitute a disposition to appropriate conduct that would make talk of rights and duties unnecessary under normal conditions. At this stage of development, however, we can analytically distinguish three major components of an Ecological Sensibility: a theory of value that recognizes intrinsic value in nature without (hopefully) engaging in more extensionism (in the sense discussed in section 3); a metaphysics that takes account of the reality and importance of relationships and systems as well as of individuals; and an ethics that includes such duties as noninterference with natural processes, resistance to human acts and policies that violate the noninterference principle, limited intervention to repair environmental damage in extreme circumstances, and a style of cohabitation that involves the knowledgeable, respectful, and restrained use of nature. Since there is not space to discuss all these components here, and since I have sketched some of them elsewhere,[3] I shall focus here on two basic dimensions of the theory of value, drawing primarily upon the writings of Leopold,[4] the Routleys,[5] and Rodman.

The first dimension is simple but sweeping in its implications. It is based upon the obligation principle that one ought not to treat with disrespect or use as a mere means anything that has a *telos* or end of its own—anything that is autonomous in the basic sense of having a capacity for internal self-direction and self-regulation. This principle is widely accepted but has been mistakenly thought (e.g., by Kant and others) to apply only to persons. Unless one engages in a high redefinition of terms, however, it more properly applies to (at least) all living natural entities and natural systems. (I leave aside in this essay the difficult and important issue of physical systems.) The vision of a world composed of many things and many kinds of things, all having their own *tele*, goes back (except for the recognition of ecosystems) to Aristotle's metaphysics and philosophy of nature and does therefore not involve us in the kinds of problems that arise from extending the categories of modern Liberal ethics to a natural world made up of the dead "objects" of modern thought. (To mention Aristotle is not, of course, to embrace all of his opinions, especially the very anthropocentric *obiter dicta*—e.g., that plants exist for the sake of animals, animals for humans, etc.—that can be found in his *Ethics* and *Politics*.) This notion of natural entities and natural systems as having intrinsic value in the specific and basic form, of having *tele* of their own, having their own characteristic patterns of behavior, their own stages of development, their own business (so to speak), is the basic ground in which is rooted the attitude of respect, the obligation of noninterference, etc. In it is rooted also the indictment of the Resource Conservation standpoint as being, at bottom, an ideology of human chauvinism and human imperialism.

It may be objected that our paradigmatic notion of a being having a *telos* is an individual human being or a person, so that viewing nature in terms of *tele* involves merely another extension of an all-too-human quality to (part of) nature, retaining the conventional atomistic metaphysics and reinstating the conventional moral pecking order. I do not think that this is the case. It seems to me an observable fact that thistles, oak trees, and wombats, as well as rainforests and chaparral communities, have their own characteristic structures and potentialities to unfold, and that it is as easy to see this in them as it is in humans, if we will but look.

For those unaccustomed to looking, Aldo Leopold's *Sand County Almanac* provides, in effect, a guidebook. Before the reader is introduced to the "land ethic" chapter (which is too often read out of the context of the book as a whole), (s)he is invited to accompany Leopold as he follows the tracks of the skunk in the January snow, wondering where the skunk is heading and why; speculating on the different meanings of a winter thaw for the mouse whose snow burrow has collapsed and for the owl who has just made dinner of the mouse; trying to understand the honking of the geese as they circle the pond; and wondering what the world must look like to a muskrat eye-deep in the swamp. By the time one reaches Leopold's discussion of the land ethic, one has grown accustomed to thinking of different animals—and (arguably), by extension, different natural entities in general—as subjects rather than objects, as beings that have their own purposes, their own perspectives on the world, and their own goods that are differentially affected by events. While we can never get inside a muskrat's head and know exactly what the world looks like from that angle, we can be pretty certain that the view is different from ours. What melts away as we become intrigued with this plurality of perspectives is the assumption that any one of them (for example, ours) is privileged. So we are receptive when the "land ethic" chapter suggests that other natural beings deserve respect and should be treated as if they had a "right" in the most basic sense of being entitled to continue existing in a natural state. To want from Leopold a full-scale theory of the rights of nature, however, would be to miss the point, since the idea of rights has only a limited application. Moreover, Leopold does not present logical arguments for the land ethic in general, because such arguments could not persuade anyone who still looked at nature as if it were comprised of objects or mere resources, and such arguments are unnecessary for those who have come to perceive nature as composed of subjects. When perception is sufficiently changed, respectful types of conduct seem "natural," and one does not have to belabor them in the language of rights and duties. Here, finally, we reach the point of "paradigm change."[6] What brings it about is not exhortation, threat, or logic, but a rebirth of the sense of wonder that in ancient times gave rise to philosophers but is now more often found among field naturalists.

In further response to the objection that viewing nature in terms of *tele* is simply another version of anthropocentric Moral Extensionism, consider that a forest may be in some ways more nearly paradigmatic than an individual human for illustrating what it means to have a *telos*. A tropical rainforest may take 500 years to develop to maturity and may then maintain a dynamic, steady-state indefinitely (for millions of years, judging from fossils) if not seriously interfered with. It exhibits a power of self-regulation that may have been shared to some extent by millennia of hunter-gatherer societies but is not an outstanding characteristic of modern humans, taken either as individuals or as societies. While there may therefore be some differences in the degree to which certain aspects of what it means to have a *telos* are present in one organism or one system compared with another, the basic principle is that all items having a *telos* are entitled to respectful treatment. Comparisons are more fruitfully made in terms of the second dimension of the theory of value.

The second dimension incorporates a cluster of value-giving characteristics that apply both to natural entities and (even more) to natural systems: diversity, complexity, integrity, harmony, stability, scarcity, etc. While the *telos* principle serves primarily to provide a common basic reason for respectful treatment of natural entities and natural systems (ruling out certain types of exploitative acts on deontological grounds), and to provide a criterion for drawing morally relevant distinctions between natural trees and plastic trees, natural forests and timber plantations, etc., this cluster of value-giving qualities provides criteria for evaluating alternative courses of permissible action in terms of optimizing the production of good effects, the better action being the one that optimizes the qualities taken as an interdependent, mutually constraining cluster. Aldo Leopold seems to have had something like this model in mind when he stated the land ethic in these terms:

A thing is right when it tends to preserve the integrity, stability, and beauty of the biotic community. It is wrong when it tends otherwise.

(We may wish to modify Leopold's statement, omitting reference to beauty and adding additional criteria, especially diversity [which stands as a principle in its own right, not merely as a means to stability]; moreover, an action can be right or wrong in reference to individuals as well as communities—but Leopold was redressing an imbalance, calling our attention to the supra-individual level, and can be forgiven for overstating his case.) More controversially, the cluster of ecological values can also be used to appraise the relative value of different ecosystems when priorities must be set (given limits on time, energy, and political influence) by environmentalists working to protect nature against the bulldozer and the chain saw. The criteria of diversity, complexity, etc., will sometimes suggest different priorities than would result from following the aesthetic of the sublime or a criterion such as sentience, while a fully pantheistic philosophy of preservation provides no criteria at all for discriminating cases.

What can be said in justification of this cluster of ecological values? It is possible for human beings to hold such values. Those who do not, and those who are not sure whether they do or not, may wish to imagine alternative worlds, asking whether they prefer the diverse world to the monocultural world, and so forth. But it would be naive to assume that such thought experiments are conducted without any significant context. For example, I am aware that my preference for diverse, complex, and stable systems occurs in a time that I perceive as marked by an unprecedentedly high rate of species extinction and ecosystem simplification. In this situation, diversity has scarcity value in addition to its intrinsic value, in addition to its instrumental value as conducive to stability. This illustrates a general characteristic of the cluster of ecological values: the balance is not static but fluctuates in response to changes in the environment, so that different principles are more or less prominent at different times.

Since the cluster of value-giving principles applies generally throughout the world to living natural entities and systems, it applies to human beings and human societies as well as to the realm of non-human nature. To the extent that diversity on an individual human level is threatened by the pressures of conformity in mass society, and diversity of social ways of life is threatened by the pressures of global resource exploitation and an ideology of world-wide "development" in whose name indigenous peoples are being exterminated along with native forests, it would be short-sighted to think of "ecological issues" as unrelated to "social issues." From an ecological point of view, one of the most striking socio-political phenomena of the twentieth century—the rise of totalitarian dictatorships that forcibly try to eliminate the natural condition of human diversity in the name of some monocultural ideal (e.g., an Aryan Europe or a classless society)—is not so much a freakish aberration from modern history as it is an intensification of the general spirit of the age. Ecological Sensibility, then, is "holistic" in a sense beyond that usually thought of: it grasps the underlying principles that manifest themselves in what are ordinarily perceived as separate "social" and "environmental" issues.[7] More than any alternative environmental ethic, it attains a degree of comprehension that frees environmentalists from the charge of ignoring "people problems" in their preoccupation with saving nature. Insofar as Ecological Sensibility transcends "ecology" in the strict sense, its very name is metaphorical, drawing on a part to suggest the whole. Starting with issues concerning human treatment of the natural environment, we arrive at principles that shed light on the total human condition.

Notes

1. The hedge ("in part") is meant to acknowledge that, although Muir did articulate the standpoint that I describe, there were other elements in his writings as well. I am analyzing a particular point of view, not presenting an exhaustive analysis of John Muir.

2. John Rodman, "Theory and Practice in the Environmental Movement," in *The Search for Absolute Values in a Changing World*, vol. 1 (New York: The International Cultural Foundation, 1978), p. 49.

3. "Ecological Resistance: John Stuart Mill and the Case of the Kentish Orchid," paper presented at the annual meeting of the American Political Science Association, 1977.

4. Aldo Leopold, *A Sand County Almanac* (New York: Oxford University Press, 1949).

5. Richard and Val Routley, "Human Chauvinism and Environmental Ethics," in *Environmental Philosophy*, eds. Don Mannison, Michael McRobbie, and Richard Routley (Department of Philosophy, Research School of Social Sciences, the Australian National University, 1980); Val and Richard Routley, "Social Theories, Self-Management, and Environmental Problems," in Ibid.

6. Obviously, I believe that those who see Leopold's land ethic as a mere extension of conventional ethics are radically mistaken.

7. See also Rodman, "Paradigm Change in Political Science." *American Behavioral Scientist* 24, 1 (September-October 1980): 67–69.

19

For a Radical Ecocentrism

Andrew McLaughlin

The need for social movements to redirect the trajectory of industrial society is urgent. An inertial tendency toward expanding the domination over the rest of nature is built into the culture and politics of industrial society. The convergence of the various ideologies of nature implicit in political economy, industrialism, and reductive science provides a powerful support for industrial culture. To an ever-increasing degree, the rest of nature is transformed into resources and commodities by industrial societies. Many want to accelerate this process.

The expansion of industrialism is honored as "economic growth," a process demanded by politicians and populace alike, appearing as if it is the only road to "prosperity." This enchantment with economic growth sprawls across most of the political spectrum, uniting conservatives, moderates, liberals, and many progressives. The political realities of industrial societies create this unity. Under conditions of economic expansion, all sides can get more of what they want, as the fruits "trickle down" even to the most disadvantaged. In times of economic contraction, struggles over the distribution of social wealth intensify. Progressives call upon the state to expand public sector expenditures meeting considerable resistance from conservatives. Since increasing social discontent threatens the control of those in power, all factions unite in calling for economic growth. But such progress is cancerous from the perspective of the rest of nature.

Ecologically sensitive progressives point to ways in which economies can be restructured, arguing that certain changes benefit both humanity and its environment. For example, retrofitting houses with insulation both generates employment and lessens the demand for energy. While this is true and, therefore, worthy of support, such win-win proposals do not get to the root problem of industrialism. The social addiction to economic growth is one of the taproots of the tension between the ecological good and industrialism, and it will remain an addiction as long as the human good is conceived as an increase in material consumption. Changing our conception of the human good involves fundamentally transforming industrialism.

Recently, some streams within the nature and social traditions are creating a significant confluence. Some within the nature tradition now recognize that creating satisfactory relations with the rest of nature requires changing society, and some within the social tradition are realizing that the creation of a just society requires resolving the ecological crises of industrial society. Each stream sees that this requires radical social change—their confluence is radical environmentalism. Radical environmentalism has five main branches: human-centered environmentalism calling for radical social change, social ecology, ecological feminism, bioregionalism, and deep ecology.[1] Human-centered radical environmentalism usually starts from the fact of ecological crisis and argues that human interests require radical social transformation.[2] This approach appeals to those who have not yet seriously questioned the assumption of anthropocentrism.[3] The limitations of such an approach become apparent when it tries to deal with questions about other species, wilderness, and the proper size of the human population. In any case, human-centered radical environmentalism is not ecocentric. Social ecology emphasizes the need to abolish society's hierarchies in order to create the possibility of ecologically harmonious relations between society and the rest of nature.[4] Ecological feminism, a part of the feminist movement, stresses the connections between the domination of women and the domination of nature.[5] It sees that at least one basic cause of the crisis is rooted in patriarchy and the masculine worldview. Deep ecology grows out of a nature tradition extending back to John Muir and Henry David Thoreau. In this essay I want to show the value of the deep ecology movement for the project of transforming industrial society. It has significant contributions to make to what I believe is a rapidly growing international radical environmental movement. I shall also briefly consider some unresolved questions which this movement must ultimately resolve.

Of these perspectives in radical environmentalism, only human-centered environmentalism accepts anthropocentrism. The others are ecocentric. Each grows out of its own history of concerns and each continues to reflect a partiality rooted in its history.[6] As the radical ecocentric movement matures, ideally the differences between these perspectives will become only differing emphases within a larger unity. There are some signs that this understanding of a deeper unity underlying diversity is developing. Marti Kheel, an ecological feminist, argues that the differences in the ways men and women now typically form their identities make a gender-neutral concept of the self suspect. This suggests that ecofeminism, in its ability to speak specifically to women and the ways that they currently form their identities, has a unique importance to women. But she sees that this does not lead to any fundamental opposition between ecological feminists and deep ecologists.[7]

Such a unity must be based on clarity about the necessity for social change. Within the United States, at least, there is a recurrent fantasy that some kind of technological invention will suffice to resolve all our problems. One could sense this reaction sweep through the United States a few years ago when it appeared that the process of "cold" fusion would offer unlimited energy in the not too distant future. Almost no one pondered whether such a development would be good; its virtue was assumed. There was almost no awareness that access to unlimited energy by this society might be catastrophic for the good of humans and certainly would be so for other forms of life. Is giving a drunk an automobile that runs on air a good idea? Even now, as we become more enmeshed in problems created by technological "solutions," we continue to yearn for the technological fix. Environmentalists are not immune to this fever.

Efforts to design with ecological sensitivity are good, but the limits of this approach are severe. Fluorescent light bulbs and superinsulated houses are design innovations that offer light and shelter with reduced environmental impact. To think that real solutions lie in this sort of tinkering is naive, dangerously so if it directs attention away from the complex, necessary, and often frustrating path of social change. Langdon Winner crisply points out that the end of appropriate technology as a social movement [in the U.S.] can be precisely located on the evening of November 4, 1980, when Ronald Reagan was elected President. His regime ended the favorable climate for appropriate technology created by the Carter administration and the movement ended.[8] The moral is that changing the direction of industrial society requires attending to the social and economic institutions of that society. Social struggle to restructure society must be part of the solution.

It should also be said, but perhaps not emphasized, that the changes needed are *radical*. Without discussing the Earth's ills at length, it needs to be said that we, all of us, face very profound problems. We have a world that threatens significant climate change with attendant mass migrations of people in a nuclearized and politically unstable system of nation states. Massive species extinction is in progress now and will continue. Agricultural lands are being turned into deserts. We have done this with somewhat over five billion people. One-quarter of those people have life styles that are vastly more destructive to the ecology than the other three-quarters. The same one-quarter is responsible for two-thirds of the global warming and almost the entire threat to the ozone layer. If we had a stable population, the spread of the life style of the rich to the poor would create a great additional strain on the environment. But we do not have a stable population. The United Nations now projects population stabilization at around eleven billion.[9] "For today's rich-country consumption levels to be achieved by a whole world that size would mean multiplying today's ecological impacts some 20 or 30 times over."[10] This simply cannot come to pass—some sort of break with our recent past is on its way.

Radical social change does not come about all at once. Michael Harrington's phrase "visionary gradualism" aptly characterizes the perspective on social changes that we need.[11] Social change must be guided by a vision of a place truly worth creating, but the changes only come about gradually, step-by-step, and require steady effort. The relevant time frame is not a few years or a decade, but generations. We should understand ourselves as living in a "transitional epoch" of several generations. A long view helps us know what should be done now.

What Deep Ecology Offers Social Progressives

The success of deep ecology or any other social movement aimed at restructuring humanity's relations with the rest of nature depends on the latent dissatisfaction of life in industrial society. If almost everyone were getting happier and happier as a result of accumulating more and more things, then calls for radical change would have little prospect of being widely heeded. In fact, as discussed, people within industrial societies do not experience increasing levels of satisfaction as they accumulate more. People become addicted to getting new things, entranced with the process of acquiring. When economies contract, people become angry, scared, insecure, and nervous about any limitation on their consumption. Industrial life swings through moods of excitement, boredom, anger, and fear. It is not a recipe for human joy or excellence.

This pattern of experience endures, not because the pattern is inherently satisfactory, but because people do not notice it. Certainly there is an awareness of the series of individual experiences, but the larger pattern exists mainly on an unconscious level. It is as if a trance has been induced by mass media, and people aimlessly graze the malls armed with credit cards, seeking *something*, though they know not what. Lacking a clear vision of a better alternative, they continue down the consumerist path. Life within industrialism is, I believe, supersaturated with latent discontent. Social progressives seem to lack a vision of the good society that does not involve spreading consumer goods around more equally. The collapse of so-called socialism into a form of market economy has not helped in envisioning an alternative to the ethos of capitalism. Although it is hard now to discern what will arise out of Eastern Europe and the dissolution of the Soviet Union, it seems that socialism, of the state capitalist variety, will be discarded in favor of the dream of shopping malls with parking lots big enough for all to come.

Perhaps, then, the most distinctive and important contribution of deep ecology to the prospect of radical change is its vision of a joyful alternative to consumerism. Environmentalism often casts itself as doomsayer, with nature as the avenging angel for industrialism's excesses. From the perspective of the dutiful consumer, this version of environmentalism speaks only of deprivation and loss. Deep ecology, on the other hand, is explicit in offering a vision of an alternative way of living that is joyous and enlivening.

This seems paradoxical, because deep ecologists have been the strongest critics of anthropocentrism, so much so that they have often been accused of a mean-spirited misanthropy. Although some who accept the deep ecology platform may have such a misanthropic streak, deep ecology is actually vitally concerned with humans realizing their best potential.[12] Arne Naess remarks, near the beginning of his major work on deep ecology [Ecosophy T], that his "discussion of the environmental crisis is motivated by the unrealized potential human beings have for varied experience in and of nature: the crisis contributes or could contribute to open our minds to sources of meaningful life which have largely gone unnoticed or have been depreciated."[13] Even though Naess's Ecosophy T recommends that people transcend their isolation from the natural world and identify with the rest of nature, his view is, in some sense, centered on humans. The reason for this apparent paradox is that deep ecology is a recommendation about how *humans* should live, and it recommends that they identify with all life, that they live nonanthropocentrically. "The change of consciousness referred to consists of a transition to a more egalitarian attitude to life and the unfolding of life on Earth. This transition opens the doors to a richer and more satisfying life for the species *Homo sapiens*, but not by focusing on *Homo sapiens*."[14] Deep ecology's conviction that this is a path toward a more joyful existence is important for any movement that would change society. "We can hope that *the ecological movement will be more of a renewing and joy-creating movement*."[15] This emphasis changes the fundamental message of radical ecocentrism from deprivation to one of a more satisfying way of being fully human.

A second major contribution of the deep ecology platform and any ecosophy that leads to it is the critique of human needs. Certainly deep ecologists are in the company of many other social theorists who distinguish between vital human needs and wants of lesser or trivial importance. One of the moral insights of social progressives is the evil of a society in which some have much and others have so little. Although the rich and the poor each have unmet needs, there is a fundamental difference between the relatively trivial dissatisfactions of the rich and the absolute deprivations of the poor. This sense of injustice, arising from an identification with the suffering of the downtrodden, was a strong element in the now dormant socialist project. "With an emphasis on this perspective, environmentalists might restore the cutting edge to socialism, restoring a sense of the absolute to perceptions of deprivation."[16]

A third contribution is deep ecology's [as Ecosophy T] focus on the process of identification. This is strategically critical in unraveling the knot of consumptive materialism.... One of the driving forces of consumerism is the loss of traditional ways of forming one's identity and their replacement by material possessions. Some sense of self-identity is a vital need for humans. Deep ecology's emphasis on an alternative mode of creating one's identity through an expansion of identification goes to the very core of one of the engines driving industrialism. An expansion of personal identification to *all* humans is a basis for rejecting consumerism in a world of desperately unmet human needs. It also leads to rejecting militarism.

Deep ecology's [as e.g. Ecosophy T] emphasis on the further expansion of identification carries an experientially based rejection of consumptive materialism. The message is not that we must "give up" some of what we want. Rather, the claim is that consumptive materialism could be lost without significant reduction in human happiness. The aim is a good *quality* of life, which is not equivalent to material consumption. The dominance of economic ways of talking and thinking about happiness fosters the illusion that consumption and happiness are equivalent. The experience of an expanded identification, along with some critical reflection, shows that they are not. Such an expanded identification is perhaps more akin to an increase in vitality and sensitivity. Increased sensitivity is a more vital way of being, but it carries a price. If one identifies with the possibly never seen rainforest, then a mahogany table, for example, beautiful in texture, grain, and craftsmanship, becomes ugly and offensive as a part of the rainforest wherein other peoples and a multitude of species once abided. Aldo Leopold is correct when he notes that "One of the penalties of an ecological education is one lives ... in a world of wounds."[17] Social progressives know a similar pain from their identification with the suffering of the poor of the world.

The expansion of identification and the ensuing reduction of the urge to consume would be helpful in alleviating some of the suffering in the Third World. If one accepts the reasonable hypothesis that overconsumption and militarism are two of the fundamental causes of the degradation of the environment, then deep ecology provides a most fundamental critique of each of these. As David Johns points out, the very concept of overconsumption changes when it is placed within the context of the deep ecology movement. Within any human-centered social perspective, overconsumption "is primarily seen as a social relationship, a problem of distribution between the wealthy and the poor, a problem of economic ownership." On the other hand, when nonhuman nature is taken as a community and valued for its own unfolding, "then human consumption which disrupts it is wrong; it would constitute overconsumption."[18] Stemming the consumerist impulse in industrial countries would slow the transfer of wealth— both in the form of material and in human life for cheap labor— from the poor of the Third World to industrial nations. If industrialism were to slow down and reverse itself, there would be much more economic and cultural room for countries of the Third World to find their own unique ways of unfolding.

Finally, the first plank of the platform calls for the flourishing of all life, human and nonhuman. If the idea of letting *every thing* flourish is powerful when applied to nature, it is even more so applied to human societies—let *every* person flourish. This fundamental norm is continuous with the ideals of all progressive social movements, going beyond them by including nonhuman life. Implicit in this ideal is the goodness of each human developing to his or her fullest. It would be hard to overestimate the impact of this ideal of human development in progressive social change. It is at the very root of political democracy and calls for economic democracy. Socialism and anarchism stand resolutely for the right of each person to realize their best potential. So then, what is wrong with capitalism? Would there be anything wrong with it if it were successful in bringing home the bacon for everyone? Yes, at least from a Marxist perspective. Capitalism promotes a systematic and deepening inequality, both economically and, more profoundly, in terms of human dignity. For Marx, the increase of wages for the worker is "nothing but *better payment for the slave,* and would not conquer either for the worker or for labour their human status and dignity."[19] Naess, too, sees the evils of economic class, noting that most of the people in industrial countries are a "global upper class" and that the core of "class suppression" is the repression "of life fulfillment potentials in relation to fellow beings."[20] Imperialism and colonialism promote a systematic inequality among humans of differing countries and inequality within each of those countries. Racism promotes a systematic inequality of equals. Sexism promotes a relationship of domination and inequality between men and women. And notice that in each of these—capitalism, racism, and sexism— the "victors" are degraded in their "victory." The racist and the sexist lose their full human potential as they enact the role of oppressor. So too with humanity and nature. The domination of nature presumes the moral irrelevance of the rest of nature, and this "victory" diminishes humans. . . .

Problems for Deep Ecology

Two criticisms that some have made of deep ecology are that it counsels passivity and that it proposes a spiritual basis for radical ecocentrism. The first charge is simply wrong, as deep ecology is anything but passive in theory or practice. In fact, it is a platform that unites many radical ecocentrists in their numerous and various struggles. Deep ecologists have often encouraged and been active in a wide range of environmental struggles. Bill Devall's recent book, for example, counsels a broad spectrum of activities as ways of "practicing deep ecology." He mentions, among others, the following actions for experiencing one's bioregion: criteria to guide choices about what one consumes; use of Naess's slogan, "simple in means, rich in ends," to focus attention on quality of life; silence as a practice; rituals; councils of all beings; intentional communities; direct action, including monkey-wrenching and "ecotage," as well as guidelines for taking direct action; political support for the politically oppressed peoples of the Third World; participation in Green movements; and much more. In fact, Devall claims that "a basic thrust of the deep, long-range ecology movement is transformation of the masses into a new kind of society."[21]

But there is still a problem. Some social activists are uneasy about deep ecology's so-called spiritual approach to social change.[22] To some extent, this charge is obviated when the distinction between the platform and the various justifications for it is emphasized. The platform is no more "spiritual" than other programs for change. On the other hand, some deep ecology writers have focused much attention on spiritual and psychological transformation. Devall, for instance, claims that the movement is "primarily a spiritual-religious movement."[23] Warwick Fox focuses his analysis on the psychological dimension of deep ecology, discussing several forms of identification without relating them to social and political questions.[24]

The problems with spiritual or psychological approaches to social change are twofold. On the one hand, they risk becoming sectarian in activity and expression, drawing lines around the "true believers," and excluding "heretics." On the other hand, it simply defies credibility to think that a spiritual or psychological foundation can support a social movement of sufficient weight to transform society. For ordinary religious movements, this limitation is not fundamental, as their primary concern is with other worlds, not the social transformation of this world. But the deep ecology movement does aim to transform society, and it is not clear that spiritual conversions or transformational psychologies are foundations that can effectively support this goal in a broadly based social movement.

This uneasiness is strengthened by the fact that there has been relatively little emphasis on social issues by deep ecologists. Sylvan is fair in claiming that the deep ecology movement has not yet developed an adequate political theory.[25] To some extent, this neglect of social issues can be attributed to the personal interests of deep ecology's founders, rather than to any deficit in deep ecology itself. It can also be explained by the relative youth of the movement whose first task has been to provide a rationale for a deep and joyful concern for all of nature. The updating and publication in English of Naess' book on deep ecology is a step toward clarifying some of the social and political dimensions of the deep ecology movement. Further developments along these lines can be expected.

As to the danger of sectarianism, many deep ecology writers explicitly attempt to be inclusive in presenting their views. This attempt, especially if emphasized, can be a check on the tendency, by no means exclusive to deep ecology, toward sectarianism. Naess is emphatic about the need to approach questions nondogmatically and to avoid sectarianism. *"Devaluation of each other's efforts within the total movement is an evil which must be avoided at all costs. No sectarianism!"*[26] Humans have a tendency to define themselves in opposition to something else. The devil is a useful persona for the true believer. This tendency seems to be exacerbated within oppositional movements, and the tendency of leftists to beat each other up in scholastic bickering is extreme. Although deep ecologists no doubt share this tendency with others, at least they have a theoretical basis for a self-definition that is inclusive instead of oppositional. Their positive valuation of diversity, both human and nonhuman, is explicit. This is a useful check on the tendency to needlessly create enemies—there are already enough real ones. . . .

The Problem of Agency

The requirement of agency derives from Marx's critique of utopian socialism. He criticized attempts at social change that appeal to "society at large" rather than to the proletariat, who had a definite interest in overthrowing capitalism.[27] Marx may have been wrong in his hope that the industrial proletariat would be the agent to transform capitalism, but his attention to the question of agency is important. The urgent question now is, who will dare to make radical ecocentric change? Insistence on this question is a contribution that theorists of social change can make to radical ecocentrism. This is just the sort of consideration that a social activist would continuously ponder, asking: Who would be my ally in this or that struggle? How can we make the movement broader and deeper? Who will act to create a solution? The problem of agency has not been a focus of deep ecology or bioregionalism, but it is arguably the "fundamental political question of the balance of this century and beyond."[28] How will such changes come about? Failure to focus on this question has been a deficit of the nature tradition. Failure to find an answer will be our common tragedy.

Perhaps there is no effective social agency for the needed transformations. Such a conclusion has practical implications. At least some activists in the Norwegian ecophilosophy movement have become convinced that industrial society cannot be effectively reformed, and that it will last only a few more decades. Therefore they decided to redirect their efforts. "After a few years, we stopped having as our aim the diversion of IGS [Industrial Growth Society] onto a socio-ecologically sound track. Instead, we started investing our activist energy into inspiring as many people as possible to experiment with a basis for a viable society to replace the one that is now step by step cracking up at its base."[29]

If there is no effective human agency, then the agent of change is the rest of nature. Naess gently suggests this when he says that "significant deterioration of ecological conditions may well color the next few years in spite of the deepening ecological consciousness. The situation has to get worse before it gets better."[30] The sense that it may not be possible to find a social agent is not restricted to deep ecologists, Earth First!, or Norway. Herman Daly, a senior economist at the World Bank and an outstanding theorist of steady-state economics, in discussing how a transition to such a steady-state economy might occur, says:

It will probably take a Great Ecological Spasm to convince people that something is wrong with an economic theory that denies the very possibility of an economy exceeding its optimal scale. But even in that unhappy event, it is still necessary to have an alternative vision ready to present when crisis conditions provide a receptive public. Crisis conditions by themselves, however, will not provide a receptive public unless there is a spiritual basis providing moral resources for taking purposive action.[31]

Unfortunately for us, there may be no effective social agency for the transformation of industrialism.

But there are some reasons for hope. Morality might deepen, and its causality could become more effective than is now apparent. I remember driving west out of Philadelphia through the normal cluttered honky-tonk atmosphere of many American highways: car lots, fried chicken stores, neon lights offering cheap gasoline, liquor stores, and supermarkets housed in cinder block buildings with no aesthetic connection to place. Place itself had been displaced. This scene repeats itself endlessly all over North America. But coming over a rise and descending into a valley, it was suddenly different. Spread out below were green fields coming forth with crops, well-kept homes and barns sparsely spread among the fields, and some people in a cart drawn by a horse. I had unexpectedly wandered into Amish country, and it had a beauty that, I imagine, was common in the last century. I was seeing a healthy agricultural community of small homesteads. The Amish, having a religiously grounded moral unity, preserve their own way of life, despite the pressures and enticements of the industrial society which surrounds them. With such unity, the agency of morality is great indeed.

If morality can be effective enough to allow the Amish to flourish in their way of life, then there may be significant parallels between the Abolitionist movement and radical ecocentrism. After all, despite the complexities of the history leading up to the Civil War, the Abolitionist movement is an example of the efficacy of moral agency. Roderick Nash, in exploring parallels between the Abolitionist movement and radical environmentalism, notes some significant similarities and draws a hopeful conclusion.[32] Political compromise was too limited for abolitionism, as it is for radical environmentalism. "Moral suasion" helped the Abolitionists, as it helps the radical environmentalists. The separatist option is limited because it leaves the problem behind. What was finally effective for the Abolitionist movement was, as we know, coercion. This is becoming part of the environmental movement as well, with the liberation of animals from laboratories, civil disobedience, Gandhian-style campaigns to save old-growth forests, and various forms of "ecotage" against the machines that would open the wilds to loggers.[33] And the state has responded with undercover agents, arrests, and charges of "conspiracy," a pattern familiar to activists of the 1960s. Some radical ecocentrists have not only been bombed, they have been arrested. As Nash points out, the possibility of freedom for slaves once seemed as remote as an ecocentric society does today. Surely the recent changes in Eastern Europe and what was once the Soviet Union indicate that major social change, even if unlikely, can happen—and rapidly.

Radical ecocentrism intends fundamental changes both in the consciousness of industrial peoples and the economic structures of industrial society. The intention toward such transformation does not make it untenable as a foundation for a broad-based social movement, but it does raise the question of how consciousness changes. Marx recognized the necessity of fundamentally transforming the consciousness of the working class in order to realize communism. Marxism approaches this problem dialectically, insisting on a constant interrelation between theory and practice and anticipating a transformation of the working class through struggle against the capitalistic class. Struggle and new forms of society lead to new forms of consciousness. In this way, Marxism offers insight into the ways in which experience leads to changes in the working-class consciousness, which in turn increase the effectiveness of the working class in its struggle against capital. This increased effectiveness leads to further social change, setting conditions for further transformations of workers, and so on. The problem for radical ecocentrism is to identify the kinds of people within the present social order who might be likely to effect the needed changes.

How can the needed consciousness arise? Clearly it does arise sometimes in some people, but it is not always clear why, even to the person affected. John Cobb remarks that "it is an exaggeration to say that I chose to become an environmentalist. On the contrary, as I became aware of the situation in new ways, I discovered that I had become an environmentalist."[34] The prospects for fundamental change exist only if such forms of consciousness become more general. But how can this happen?

If there is a developmental dialectic between the environmental activism and radical ecological consciousness, such that an increase in one tends to foster growth in the other, then there is hope for the future effectiveness of radical ecocentrism. It seems almost certain that more and more people will become aware of increasingly severe environmental problems, if only because they are personally touched by them. At least some people will act on that awareness through various forms of social protest. As citizens become involved in environmental problems of direct concern to them, they find that other issues are connected and also require attention. Resistance to a proposed incinerator or concern about toxic wastes in the community's drinking water can lead rather directly to larger questions about the production of garbage or industrial processes that generate toxic wastes. Such concerns can easily develop into doubts about the larger production and distribution systems that generate both garbage and hazardous wastes. It is a small step from here to a view of the whole mode of industrial production as the problem. This opens the door to the radical ecocentric platform. Of course not every person who becomes involved in a local problem will be led to radical ecocentrism, but the potential is there. People can make connections between their problems and larger issues because the connections are really there.

Similarly, workers' concerns about their health lead unions to demand the right to know what their members are exposed to on the job. They might then demand the right to refuse to work, without the loss of wages, when workers would be exposed to hazardous substances. Community activists might organize for similar rights to know and refuse in their community. When labor *and* community activists lend support to each other on these issues, there arises a powerful basis for effective movements to confront and change current modes of production.[35]

Interestingly, such movements can be particularly effective when they follow a two-pronged strategy of political action to force governmental regulation of hazardous substances and community action to compel the enforcement of these regulations.[36] This strategy leads to the possibility that the contested action might be prevented entirely. The fact that there are problems in getting approval for incinerators, dump sites, and nuclear power stations exemplifies the effectiveness of a combination of governmental regulations—necessitated by environmental political pressure—and community activism at the proposed sites. As Andrew Szasz notes, such activism "pits people against capital and state regulators, provides the empowering experience of collective action, and radicalizes participants."[37] Such a model of the dialectic between activism and consciousness gives grounds for hopes of success both in specific struggles and in the spread of a radical ecocentric consciousness—not certainty, but hope.

On a more national scale, the Green movement within the United States is a seed of radical ecocentrism which might blossom into a significant social and political force. Their recently adopted program spans many areas, including positions and policies on arts and education, economics, direct action, energy, agriculture, and biological diversity.[38] This program was only developed after long discussion within local chapters and between those chapters, and it reflects consensus in the movement. Throughout the program, ecocentric values are affirmed, including the respect for nature, plants, animals, and species diversity, and there are calls for the expansion of wilderness and the preservation of native cultures. These values are combined with an understanding of the bankruptcy of both capitalistic and state socialistic regimes and an advocacy of decentralized and regionalized economies. They call for public control of banking and energy but intend such control to be held by a "decentralized public sector." Such public control would not prohibit individuals, small cooperatives, and small companies from free economic activity. Rather it seeks to resubmerge local economies into a social context that puts human development and ecology before profit. Thus the

Greens seek to reverse the dominance of the economy over all other aspects of society, which is a central trait of industrialism. The emergence of the Green movement in the United States will, no doubt, continue to face difficulties similar to those of Europe, but there is hope in the fact that such movements have achieved the degree of success they have in what is, in proper perspective, a very short time.

The importance of feminism in helping to create the social agency for change is vast. As Ynestra King argues, "potentially, feminism creates a concrete global community of interests among particularly life-oriented people of the world: women."[39] However, as she notes, feminism does not, necessarily, lead to an ecological feminism. It is possible for feminism to accept the project of domination over nature, which is, she says, the position of many socialist-feminists. What is needed is a specifically ecological feminism that can "integrate intuitive, spiritual, and rational forms of knowledge, embracing both science and magic insofar as they enable us to transform the nature-culture distinction and to envision and create a free, ecological society."[40] If ecological feminists were to be effective in shifting the direction of feminism in an ecological direction, the hopes for an effective radical ecocentrism would increase enormously.

Anticipating the development of radical ecocentrism into a major political force, we should note three contributions from radical social theory. First, the value of equality within society is extremely important, both as an element of social justice and as an antidote to the social demand for economic expansion. If material equality becomes recognized as a social good, then the process of trying to achieve status through material possession would be weakened. In a society where having more than others is regarded as rude and tasteless, rather than as a mark of merit or status, demands for economic growth would be muted.

Second, until such social equality is attained, there will be a tendency for divisive conflict between social progressives and radical ecocentrists. There is a latent fracture line between social progressives and ecocentrists when environmental reform involves economic costs. In such situations, the distribution of the burdens of such reforms will be a politically important question. When such reforms seem unavoidable, the rich will seek to place their costs on the backs of the poor.[41] Social progressives will resist this, perceiving its fundamental injustice. But where will the radical ecocentrists stand? They may be tempted to endorse the reform, despite the fact that it increases social inequality, because it might abate some serious environmental problem.

To decide where they should stand when this type of immediate reform increases inequality, they must understand the real causes of the destruction of nature. If they understand that capitalism and industrialism must be undone, then the dangers of expedient alliances with the wealthy will be clear. Minor reforms are to be welcomed, to be sure, but not if they are carried out at the expense of the poor. To accede to such socially regressive reforms, even if environmentally desirable, corrodes the possibility of political alliances between radical ecocentrists and social progressives. If radical ecocentrists do not stand consistently on the side of the poor, they will lose one of their major allies for the changes that *really* need to be made. Only a clear understanding of industrialism's determinative role in the rape of the Earth can foster resistance to this division. To change industrialism, radical ecocentrism must make common cause with the oppressed of the world.

Finally, just as humanity and nature need to be understood within a holistic perspective, so too does modern social reality require a holistic understanding.[42] The various forms of oppression include economic exploitation, racism, sexism, patriarchy, heterosexism, nationalism, authoritarianism, and the domination over nature; each exists in overt and subtle forms. Each of these forms of oppression requires struggle, but struggle against only one probably cannot be effective. The systemic integration of modern society makes them intertwined. Concern for any oppression requires concern for all oppression. This implies an objective basis for a broad-based political unity, and in this there is hope.

The real problem is the behemoth of industrialism ever expanding its grip, around the globe, beneath the earth, and into the skies. The seeds of an effective and radical ecocentrism live in those who somehow awaken to the exhilaration of being human in harmony with the rest of nature. Some may choose to stand for the forests, and that is good. Others, however, must reach out to the oppressed of the world and build bridges between the poor who live within the industrial world and those at the periphery. This requires political organization and action. It is another essential path.

Although grounds for despair are all around, there are also signs of hope. The environmental movement has broadened and deepened with astonishing rapidity. The Green movement is a significant presence in the national politics of many countries. Environmental groups supporting direct action, such as Earth First!, Sea Shepherd, and Greenpeace often have significant support from the general public. Feminism, both in theory and in practice, has become an important force for social change, and ecological feminism will probably increase in importance within feminism.

It is hard to know now whether such movements can gain sufficient strength in numbers to force a reversal of the structure of industrialism. We do not know how much time we have. We may be in a transitional epoch and not the closing hours of industrialism—no one really knows. If there is time, then building an effective radical ecocentric movement founded on the desire for community life, on a rejection of social domination and sexual oppression, on an empathy for other animals, and on a love for all of nature may be able to effectively challenge and restructure industrialism.

We have recently taken a few steps back from the nuclear abyss. Should demilitarism spread, as it can if citizens continue to press for a real peace, this will release massive social resources for social and ecological reconstruction. Numerous projects must be funded, including reforestation, stopping desertification, controlling and reversing population growth, developing energy efficiency and eventually ceasing the use of fossil fuels, stopping the loss of rainforests, expanding wilderness, ending the human extinction of animal and plant species, protecting the ozone shield, dealing with climate change, and many other tasks. Since the world military budget is some $900 billion each year, the financial resources are there if we can demilitarize society. The point is not that it is easy, but that it is possible. Peace action remains critical.

The struggle will extend beyond any of our lifetimes, and it is important to live now in a way that enables one's spirit to flourish. Humility, humor, and compassion are necessary to stay on the path. It is important to remember that mistakes are part of the way and results are often ambiguous. While the search for purity is admirable, its attainment is impossible and the fruits of action are uncertain. The point is the action, not its fruit. Such an understanding can sustain us through the hard times with a joy in all existence and an appreciation of our fellow travelers. As Norman Cousins put it, none of us "knows enough to be a pessimist."[43]

Notes

1. The best comprehensive discussion of ecofeminism, social ecology, bioregionalism, and ecocentric political theory in general is in Robyn Eckersley's *Environmentalism and Political Theory: Toward an Ecocentric Approach* (Albany: State University of New York Press, 1992).

2. For an excellent anthropocentric radical environmentalism, see Lester Milbraith, *Envisioning a Sustainable Society: Learning Our Way Out* (Albany: State University of New York Press, 1989). See also John Dryzek, *Rational Ecology: Environment and Political Economy* (New York: Basil Blackwell, 1987), Joel Jay Kassiola, *The Death of Industrial Civilization: The Limits to Economic Growth and the Repoliticization of Advanced Industrial Society* (Albany: State University of New York Press, 1990), and William Ophuls, *Ecology and the Politics of Scarcity* (San Francisco: W.H. Freeman & Co., 1977).

3. I do not mean to characterize the beliefs of these authors, as they do not directly address the question of anthropocentrism. Dryzek uses an anthropocentric perspective, but he does not claim that it is adequate. He uses it as a minimal assumption and adopts it because such an assumption is embedded in the perspectives which he wishes to meet on their own ground, *Rational Ecology*, 35.

4. Murray Bookchin has generated numerous books and articles in his long and consistent advocacy of social ecology, an ecologically informed variant of anarchism. The main work is *The Ecology of Freedom: The Emergence and Dissolution of Hierarchy* (Palo Alto: Cheshire Books, 1982). His most recent statements of his position are *Remaking*

Society: Pathways to a Green Future (Boston: South End Press, 1990) and *The Philosophy of Social Ecology: Essays on Dialectical Naturalism* (Montreal: Black Rose Books, 1990).

5. A good discussion, though not easily accessible, is Val Plumwood, "Ecofeminism: An Overview and Discussion of Positions and Arguments," *Australasian Journal of Philosophy*, Supplement to vol. 64 (June 1986): 120–138. A less theoretical but lively expression of ecofeminism is *Healing the Wounds: The Promise of Ecofeminism*, edited by Judith Plant (Philadelphia: New Society Publishers, 1989). See also *Hypatia* 6, 1 (Spring 1991), which is a special issue on ecological feminism.

6. The human-centered environmentalists rarely argue for their anthropocentrism. Social ecologists focus on social hierarchy, ecological feminists focus on gender-rooted causes, and deep ecologists emphasize anthropocentrism.

7. See Marti Kheel, "Ecofeminism and Deep Ecology: Reflections on Identity and Difference," *The Trumpeter* 8, 2 (Spring 1991): especially 70. See also Karen Warren's discussion of the "boundary conditions" of ecofeminist ethics. She includes "descriptions and prescriptions of social reality that do not maintain, perpetuate, or attempt to justify social 'isms of domination' and the power-over relationships used to keep them intact." "Ecological Feminism and Ecosystem Ecology," *Hypatia* 6, 1 (Spring 1991): 181. Although she does not mention it, this would clearly include deep ecology within ecofeminist ethics. I think the perception of unity will spread to most radical environmentalists as the movement develops further.

8. Langdon Winner, *The Whale and the Reactor: A Search for Limits in an Age of High Technology* (Chicago: University of Chicago Press, 1986), 80.

9. Nafis Sadik, *The State of World Population 1990* (New York: United Nations Population Fund, n.d.).

10. James Robertson, "Toward a Multi-Level One-World Economy," *New Options* 63:2.

11. Michael Harrington, *Socialism: Past and Future* (New York: Penguin Books, 1990), chap. 9.

12. Naess is clear that his view is not misanthropic. His negative reaction towards the increase of human population is not to foster any animosity towards humans as such—on the contrary, human fulfillment seems to *demand* and *need* free nature. Arne Naess, *Ecology, Com-*

munity and Lifestyle: Outline of an Ecosophy, translated and revised by David Rothenberg (New York: Cambridge University Press, 1989), 141, emphasis in original.

13. Naess, *Ecology*, 24.

14. Naess, *Ecology*, 91.

15. Naess, *Ecology*, 91, emphasis in original. He also states: "Human nature may be such that with increased maturity a *human* need increases to protect the richness and diversity of life for *its own sake*. Consequently, what is useless in a narrow way may be useful in a wider sense, namely satisfying a human need." Naess, *Ecology*, 177, emphasis in original.

16. Robert C. Paelke, *Environmentalism and the Future of Progressive Politics* (New Haven: Yale University Press, 1989), 168.

17. Aldo Leopold, *A Sand County Almanac: With Essays on Conservation from Round River* (New York: Ballantine Books, 1979), 197.

18. David M. Johns, "The Relevance of Deep Ecology to the Third World," *Environmental Ethics* 12, 3 (Fall 1990): 242.

19. Karl Marx, "Economic and Philosophic Manuscripts," in *The Marx-Engels Reader*, edited by Robert C. Tucker, 2nd ed. (New York: Norton, 1978), 80, emphasis in original. Interestingly, Marx also recognizes the process of extending identification when he claims that when workers unite in struggle, they acquire a new need for society: "The brotherhood of man is no mere phrase with them, but a fact of life." "Manuscripts," 100–101.

20. Naess, *Ecology*, 138.

21. Bill Devall, *Simple in Means, Rich in Ends: Practicing Deep Ecology* (Salt Lake City: Gibbs Smith, 1985), 128.

22. See Richard Sylvan, "A Critique of (Wild) Western Deep Ecology," unpublished manuscript, 30ff.

23. Devall, *Simple in Means*, 160.

24. See Warwick Fox, *Towards a Transpersonal Ecology: Developing New Foundations for Environmentalism* (Boston: Shambhala, 1990). I should note that Fox's concern in this book is with what is philosophically distinctive about deep ecology, 43. Although he recognizes the political dimensions of deep ecology, that is not the subject of his book.

25. Richard Sylvan, "A Critique of Deep Ecology," 39.

26. Naess, *Ecology*, 91, emphasis in original.

27. Tucker, *Marx-Engels Reader*, 498.

28. Paelke, *Environmentalism*, 203.

29. Sigmund Kvaloy, "Norwegian Ecophilosophy and Ecopolitics and Their Influence from Buddhism," in *Buddhist Perspectives on the Ecocrisis*, edited by Klas Sandell (Kandy, Sri Lanka: Buddhist Publication Society, 1987), 62. Dave Foreman, a founder of the Earth First! movement, has argued that the movement is a holding action to save species and old-growth forests until industrialism collapses.

30. Naess, *Ecology*, 211.

31. Herman E. Daly, "The Steady-State Economy: Postmodern Alternative to Growthmania," in *Spirituality and Society: Postmodern Visions*, edited by David Ray Griffin (Albany: State University of New York Press, 1988).

32. See Roderick Frazier Nash, *Rights of Nature: A History of Environmental Ethics* (Madison: University of Wisconsin Press, 1989), the epilogue.

33. See Christopher Manes, *Green Rage: Radical Environmentalism and the Unmaking of Civilization* (Boston: Little, Brown & Co., 1990) for a sympathetic and up-to-date account of the activities of radical ecocentrists.

34. John B. Cobb, Jr., "Postmodern Social Policy," in *Spirituality and Society*, Griffin, ed., 102.

35. Such was the outcome of the Third New York State Environment and Labor Conference, November 1991, which had delegates from a wide spectrum of labor and community groups.

36. See Andrew Szasz, "In Praise of Policy Luddism: Strategic Lessons from the Hazardous Waste Wars," *Capitalism, Nature, Socialism* 2, 1 (February 1991): 17–43.

37. Szasz, "In Praise of Policy Luddism," 43.

38. Copies of the program are available for $4.00 from The Greens Clearinghouse, PO Box 30208, Kansas City, MO 64112, 816–931–9366.

39. Ynestra King, "The Ecology of Feminism and the Feminism of Ecology," in *Healing the Wounds*, Judith Plant, ed., 1989.

40. King, "The Ecology of Feminism," 23.

41. Hugh Stretton, *Capitalism, Socialism, and the Environment* (New York: Cambridge University Press, 1976), 48–49.

42. See Michael Albert, Leslie Cagan, Noam Chomsky, Robin Hahnel, Mel King, Lydia Sargent, and Holly Sklar, *Liberating Theory* (Boston:

South End Press, 1986) for an attempt to articulate a holistic theory that links various forms of oppression. This book does not question its own anthropocentrism.

43. Norman Cousins in *Whole Earth Review* No. 61 (Winter 1988): 20.

Recent Books Relevant to the
Deep Ecology Movement Plus a Few Classics

* Can be read for the philosophical core.
\+ Are considered classics.

+Abbey, Edward. 1968. *Desert Solitaire*. University of Arizona Press, Tucson.

Alexander, Christopher. 1979. *The Timeless Way of Building*. Oxford University Press, New York.

Amidon, Elias & Elizabeth Roberts. 1991. *Earth Prayers*. Harper, New York.

Barrett, William. 1978. *The Illusion of Technique*. Anchor, New York.

Berger, John. 1985. *Restoring the Earth*. Knopf, New York.

*Berman, Morris. 1981. *The Reenchantment of the World*. Cornell University Press, Ithaca, NY.

Berman, Morris. 1989. *Coming to Our Senses: Body and Spirit in the Hidden History of the West*. Bantam, New York.

Berry, Thomas. 1988. *Dream of the Earth*. Sierra Club Books, San Francisco.

Berry, Wendell. 1977. *The Unsettling of America: Culture and Agriculture*. Sierra Club Books, San Francisco.

Bly, Robert. 1988. *A Little Book on the Human Shadow*. Edited by W. Booth. Harper & Row, New York.

Bookchin, Murray. 1982. *The Ecology of Freedom: The Emergence and Dissolution of Hierarchy*. Cheshire Books, Palo Alto, CA.

Bookchin, Murray & Dave Foreman. 1991. *Defending the Earth: Debate Between Murray Bookchin and Dave Foreman*. Black Rose Books, Montreal.

*Bowers, C.A. 1993. *Critical Essays on Education, Modernity, and the Recovery of the Ecological Imperative*. Teachers College Press, New York.

Cajete, Gregory. 1994. *Look to the Mountain: An Ecology of Indigenous Education*. Kivaki Press, Durango, CO.

Capra, Fritjof. 1982. *The Turning Point: Science, Society and the Rising Culture*. Simon and Schuster, New York.

Capra, Fritjof & David Steindl-Rast. 1991. *Belonging to the Universe: Explorations on the Frontiers of Science and Spirituality*. Harper Collins, New York.

+Carson, Rachel. 1962. *Silent Spring*. Houghton Mifflin, Boston.

+Commoner, Barry. 1972. *The Closing Circle*. Bantam, New York.

+Dasmann, Ray. 1966. *The Destruction of California*. Macmillan, New York.

Davis, Donald E. 1989. *Ecophilosophy: A Field Guide to the Literature*. R. & E. Miles, San Pedro, CA.

DeGroh, Teresa & Edward Valanskas. 1987. *Deep Ecology and Environmental Ethics: A Selected and Annotated Bibliography of Materials Published since 1980*. Council of Planning Librarians, Chicago.

*Devall, Bill & George Sessions. 1985. *Deep Ecology: Living as if Nature Mattered*. Gibbs Smith, Salt Lake City.

*Devall, Bill. 1988. *Simple in Means, Rich in Ends: Practicing Deep Ecology*. Gibbs Smith, Salt Lake City.

*Devall, Bill. 1993. *Living Richly in an Age of Limits*. Gibbs Smith, Salt Lake City.

*Devall, Bill (ed.). 1994. *Clearcut: The Tragedy of Industrial Forestry*. Sierra Club Books, Earth Island Press, San Francisco.

Diamond, Stanley. 1981. *In Search of the Primitive*. Transaction Books, New York.

+Dogen. 1985. *Shobogenzo*. Translated by T. Cleary. University of Hawaii Press, Honolulu.

Drengson, Alan. 1983. *Shifting Paradigms: From Technocrat to Planetary Person*. LightStar, Victoria, BC, Canada.

*Drengson, Alan. 1989. *Beyond Environmental Crisis: From Technocrat to Planetary Person*. Peter Lang, New York.

Drengson, Alan. 1993. *Doc Forest and Blue Mountain Ecostery: A Narrative on Creating Ecological Harmony in Daily Life*. Ecostery House, Victoria, BC, Canada.

*Drengson, Alan. 1995. *The Practice of Technology*. SUNY Press, Albany, NY.

Duncan, David. 1983. *The River Why*. Bantam, New York.

*Eckersley, Robyn. 1992. *Environmentalism and Political Theory: Toward an Ecocentric Approach*. SUNY Press, Albany, NY.

Ehrenfeld, David. 1978. *The Arrogance of Humanism*. Oxford University Press, New York.

Ehrenfeld, David. 1991. *Beginning Again: People and Nature in the New Millennium*. Oxford University Press, New York.

+Ehrlich, Paul. 1969. *The Population Bomb*. Sierra Club Books, San Francisco.

Ehrlich, Paul & Ann. 1981. *The Causes and Consequences of the Disappearance of Species*. Random House, New York.

Evernden, Neil (ed.). 1984. *The Paradox of Environmentalism*. A Symposium, May 2, 1983, York University. Faculty of Environmental Studies, York University, Downsview, Ontario.

Evernden, Neil. 1985. *The Natural Alien: Humankind and Environment*. University of Toronto Press, Toronto.

Evernden, Neil. 1992. *The Social Creation of Nature*. Johns Hopkins University Press, Baltimore.

Fox, Matthew. 1988. *The Coming of the Cosmic Christ*. Harper & Row, San Francisco.

*Fox, Warwick. 1990. *Toward a Transpersonal Ecology: Developing New Foundations for Environmentalism*. Shambhala, Boston.

*Goldsmith, Edward. 1993. *The Way: An Ecological Worldview*. Shambhala, Boston.

Gore, Al. 1992. *Earth in the Balance*. Houghton Mifflin, New York.

Grossinger, Richard (ed.). 1992. *Ecology and Consciousness*. North Atlantic Books, Berkeley, CA.

*Grumbine, Edward R. 1992. *Ghost Bears: Exploring the Biodiversity Crisis*. Island Press, Covelo, CA.

Hargrove, Eugene. 1989. *Foundations for Environmental Ethics*. Prentice-Hall, Englewood Cliffs, NJ.

Harner, Michael. 1986. *The Way of the Shaman*. Bantam, New York.

Hawkin, Paul. 1993. *The Ecology of Commerce*. Harper Business, New York.

Highwater, Jamake. 1981. *The Primal Mind*. Harper & Row, New York.

Kaza, Stephanie. 1993. *The Attentive Heart*. Fawcett Columbine, New York.

Kohák, Erazim. 1984. *The Embers and the Stars: A Philosophical Inquiry into the Moral Order of Nature*. University of Chicago Press, Chicago.

Kohr, Leopold. 1978. *The Overdeveloped Nations: The Diseconomies of Scale*. Schocken Books, New York.

LaChapelle, Dolores. 1978. *Earth Wisdom*. Finn Hill Arts, Silverton, CO.

*LaChapelle, Dolores. 1988. *Sacred Land, Sacred Sex: The Rapture of the Deep*. Kivaki Press, Durango, CO.

+Lao Tzu. 1963. *Tao Te Ching*. Translated by D.C. Lau. Penguin, New York.

+Leopold, Aldo. 1949. *A Sand County Almanac: And Sketches Here and There*. Oxford University Press, New York.

Livingston, John A. 1973. *One Cosmic Instant: Man's Fleeting Supremacy*. Houghton Mifflin, Boston.

Livingston, John A. 1981. *The Fallacy of Wildlife Conservation*. McClelland and Stewart, Toronto.

*Macy, Joanna. 1991. *World as Lover, World as Self*. Parallax Press, Berkeley.

*Mander, Jerry. 1991. *In the Absence of the Sacred: The Failure of Technology and the Survival of the Indian Nations*. Sierra Club Books, San Francisco.

Manes, Christopher. 1990. *Green Rage: Radical Environmentalism and the Unmaking of Civilization*. Little, Brown and Company, Boston.

Maser, Chris. 1988. *The Redesigned Forest*. R. & E. Miles, San Pedro, CA.

*Mathews, Freya. 1991. *The Ecological Self*. Barnes & Noble, Savage, MD.

*McLaughlin, Andrew. 1993. *Regarding Nature: Industrialism and Deep Ecology*. SUNY Press, Albany, NY.

+Meadows, Donella H., et al. 1972. *The Limits to Growth: A Report for the Club of Rome's Project on the Predicament of Mankind*. Universe Books, New York.

Meeker, Joseph. 1980. *The Comedy of Survival: In Search for an Environmental Ethic.* Guild of Tutors, Los Angeles.

Merchant, Carolyn. 1992. *Radical Ecology: The Search for a Livable World.* Routledge, New York.

+Muir, John. 1954. *The Wilderness World of John Muir.* Edited by E.W. Teale. Houghton Mifflin, New York.

Naess, Arne. 1974. *Gandhi and Group Conflict: Exploration of Satyagraha, Theoretical Background.* Universitetsforlaget, Oslo.

*Naess, Arne. 1989. *Ecology, Community and Lifestyle: Outline of an Ecosophy.* Translated and revised by D. Rothenberg. Cambridge University Press, Cambridge.

Norberg-Hodge, Helena. 1991. *Ancient Futures.* Sierra Club Books, San Francisco.

Oelschlaeger, Max. 1991. *The Idea of Wilderness.* Yale University Press, New Haven.

Orr, David. 1992. *Ecological Literacy: Education and the Transition to a Postmodern World.* SUNY Press, Albany, NY.

Ponting, Clive. 1991. *A Green History of the World: The Environment and the Collapse of Great Civilizations.* Penguin, New York.

Quinn, Daniel. 1993. *Ishmael.* Bantam, New York.

*Roszak, Theodore. 1978. *Person/Planet: The Creative Disintegration of Industrial Society.* Anchor Press/Doubleday, Garden City, NY.

Roszak, Theodore. 1992. *The Voice of the Earth.* Simon & Schuster, New York.

Rothenberg, David (ed.). 1992. *Wisdom of the Open Air: The Norwegian Roots of Deep Ecology.* University of Minnesota Press, Minneapolis.

*Rothenberg, David. 1993. *Is It Painful to Think? Conversations with Arne Naess.* University of Minnesota Press, Minneapolis.

Rowe, Stan. 1990. *Home Place: Essays on Ecology.* Newest, Edmonton.

+Sahlins, Marshall. 1972. *Stone Age Economics.* Aldine, New York.

Sale, Kirkpatrick. 1985. *Dwellers in the Land: The Bioregional Vision.* Sierra Club Books, San Francisco.

Seed, John, et al. (eds.) 1988. *Thinking Like a Mountain: Towards a Council of All Beings*. New Society Publishers, Philadelphia.

*Sessions, George & Bill Devall. 1985. *Deep Ecology: Living as if Nature Mattered*. Gibbs Smith, Salt Lake City.

Sessions, George (ed.). 1995. *Deep Ecology for the 21st Century*. Shambhala, Boston.

*Shepard, Paul. 1973. *The Tender Carnivore and the Sacred Game*. Charles Scribner's Sons, New York.

Shiva, Vandana. 1993. *Monocultures of the Mind*. Third World Network, Penang, Malaysia.

Skolimowski, Henryk. 1981. *Eco-Philosophy: Designing New Tactics for Living*. Marion Boyars, Boston.

Skolimowski, Henryk. 1992. *Living Philosophy: Eco-Philosophy as a Tree of Life*. Arkana, London.

Smith, Houston. 1982. *Beyond the Post-Modern Mind*. Crossroad, New York.

*Snyder, Gary. 1977. *The Old Ways: Six Essays*. City Lights Books, San Francisco.

Snyder, Gary. 1980. *The Real Work: Interviews and Talks, 1964–1979*. New Directions, New York.

*Snyder, Gary. 1990. *The Practice of the Wild*. North Point, San Francisco.

Thoreau, Henry D. 1980. *The Natural History Essays*. Gibbs Smith, Salt Lake City.

Tobias, Michael (ed.). 1985. *Deep Ecology*. Avant Books, San Diego.

*Toulmin, Stephen. 1990. *Cosmopolis: The Hidden Agenda of Modernity*. University of Chicago Press, Chicago.

Trungpa, Chögyam. 1984. *Shambhala: The Sacred Path of the Warrior*. Shambhala, Boston.

Tucker, Mary Evelyn & John A. Grim (eds.). 1993. *Worldviews and Ecology*. Bucknell University Press, Lewisburg, PA.

Wilber, Ken. 1981. *Up From Eden: A Transpersonal View of Religion*. Anchor, New York.

Wilson, Edward O. 1992. *The Diversity of Life*. Harvard University Press, Cambridge.

*Zimmerman, Michael E. 1990. *Heidegger's Confrontation with Modernity*. Indiana University Press, Bloomington.

APPENDIX

Ecoforestry Statement of Philosophy
from the Ecoforestry Institute

Introduction

The emergence of ecoforestry coincides with the deepening of the environmental crisis. The manifestations of this crisis are all around us and well known. The environmental movement has roots in earlier times, but has become a worldwide grass roots movement. There are two main forms of environmentalism, one is the *status quo, mild reform movement*. The other, the *Deep Ecology Movement*, is based on the awareness that we cannot go on with business as usual. We must make fundamental and sweeping changes in our values, philosophies and practices. From the platform of the Deep Ecology Movement, it is clear that the crisis we face is a crisis in culture, character and consciousness.

George Sessions and Arne Naess formulated a version of the platform principles for the deep ecology movement in 1984. These principles recognize that we must make cultural and individual changes based on respect for the intrinsic worth of all natural beings. Since mild reform continues to treat nature as a source of only instrumental values for human support, profit and enjoyment, it cannot end the destruction of nature.

Acceptance of the platform principles leads to new practices which respect and work *with* the values *inherent* in ecological communities, natural beings and their processes. In the case of forestry, this new practice is called *ecoforestry* to signify its break with traditional forestry. Traditional Western forestry is based on an agricultural, industrial model that views natural forests as something to be replaced with technologically designed and managed tree plantations. It is now obvious that this approach has failed, and that its continuing practice will destroy all natural forests; it plays a *major* role in global ecological destruction and species extinction. And finally, it is wrong to destroy ecological contexts, the basis of life for all beings.

Ecoforestry realizes ecologically responsible forest use by means of commitments and new practices based on *ecocentric* values. Its practices follow from accepting the platform principles of the deep ecology movement in conjunction with leading-edge knowledge of forest ecology, evolution, landscape ecology, conservation biology, stand-level practical knowledge and the vernacular wisdom found among forest *dwellers*, primarily but not only indigenous peoples. Ecoforestry is perennial forest use, based on respect for the wisdom and intrinsic values of the forest. In conjunction with the platform principles, the Ecoforestry Institute urges that we immediately take all of the necessary steps to recycle, reduce use and waste of all resources. We must find substitutes for wood fiber in most applications. Timber taken from forests must be used for high-quality products which are durable.

The Institute endorses diversified forest-based economies made up of small businesses, workers' co-ops, non-profit organizations, small corporations, private land, and land trusts. Finally, the Institute in cooperation with the SILVA Foundation and other organizations is building a certification system which informs users of wood products how to choose those obtained and manufactured by ecologically responsible standards and practices.

Because of the deep responsibilities of ecoforesters, an oath which they strive to live up to has been developed. It distills and summarizes the philosophy and values of ecoforestry practices. It is hoped that the oath stated below inspires general agreement and will be adopted by all organizations and individuals devoted to ecologically responsible forest use, whatever their activities and products.

The Ecoforesters' Way:
Oath of Ecologically Responsible Forest Use

1. We shall respect, hold sacred, and learn from the ecological wisdom (ecosophy) of natural forests with their multitudes of beings;

2. We shall protect the integrity of the full functioning forest;

3. We shall not use agricultural practices on the forest;

4. We shall remove from forests only values which are in abundance and to meet vital human needs;

5. We shall remove individual instances of values only when this removal does not interfere with full functioning forests: when in doubt, we will not;

6. We shall minimize the impacts of our actions on the forest by using only appropriate, low-impact technology practices;

7. We shall use only non-violent resistance (e.g., Gandhian methods) in our protection of the forests;

8. We shall do good work and uphold the Ecoforester's Way as a sacred duty and trust.

A version of the following opinion advertisement appeared in *The New York Times* on May 25, 1993, in support of the above oath.

A TREE FARM IS *NOT* A FOREST

On the eve of the President's new forestry plan, the right issues are still not addressed. The problem is not loggers' jobs versus spotted owls. That is the industry's equation. The REAL problem is the present concept of forestry: FORESTRY AS INDUSTRIAL AGRICULTURE. This cannot be sustained. It is already failing. Industrial forestry cannot save jobs, or communities. It kills the forests and kills our nation's heritage and spirit. That's the bad news. The good news is that wonderful alternatives exist. But radical change is required.

1. The Forest as Agriculture

The first American foresters were trained in Germany in the late 1800s. They were taught to grow trees as if they were corn. They cut down the *native* forests, and then replanted with single (or few) species of trees, usually in long rows. They harvested them all when they reached the same age, and replanted again.

We now have a name for this, *industrial forestry*: trees as assembly line parts, trees viewed strictly as products, as "board feet," as digits in the GNP, as *cash crops*. But forests are not agriculture. Viewing them this way is the root of our problem.

Natural forests are ecological communities. They contain thousands of life forms, including diverse species of trees, animals, insects, plants and microorganisms, all at different stages in their evolutionary cycles, all in constant interaction, co-evolving with one another, with water and fire, wind and avalanche. *A forest is dynamic*: one of nature's great cauldrons of biological diversity and genetic richness. Only by respecting and preserving that diversity can forests be healthy. And only if they are healthy, in the long term, can we be healthy.

Natural forests are self-sustaining, and self-repairing. They do not need humans. But humans can have a role, and we benefit greatly from healthy forests. All benefit ceases when forests are misunderstood, depleted, simplified and redesigned for human purposes.

2. The Failures of Industrial Forestry

Industrial forestry—tree farms, plantations, clearcutting—*kills* forests, wiping out thousands of years of genetic heritage. When the rich undergrowth of shrubs and grasses are cleaned out, when diverse tree species are eliminated, when the tussock moths and bark beetles (and spotted owls) lose their habitat, when the bugs and worms that feed the birds are gone, then the forest is gone. *Ten million newly planted sprigs of pine is not a forest.*

Tree plantations weaken the ecosystem. They cannot cope with many of nature's challenges. They are dependent on humans the way caged monkeys or rows of tomato plants are. When miles of diverse natural forest are reduced to only one species, the entire crop is vulnerable to blight. *Heavy pesticide use is required to protect it.*

And when the biologically rich soup that is a natural forest is simplified, the soil is no longer enriched by the nutrients it needs; it too becomes weakened and unproductive. *Fertilizers are required to keep the soil productive.* And *then*, when the fertilizers and pesticides find their way into streams, the water is poisoned, the fish die, and entire ecosystems are threatened. Ecological collapse is around the corner, with grave consequences to humans, too.

As for *clearcutting*—industrial forestry's most spectacular statement—this eliminates much more than just trees and forests; it often removes the land itself. With no roots to hold the soil, whole hillsides slump into streams, where the rich soils become useless and sometimes lethal to life.

Clearcutting eliminates something else, as well. *Jobs.* When the trees are gone, so are the jobs. In many parts of this country, that outcome is already apparent. Clearcutting and plantation forestry have destroyed the forests and ecological vitality, while also bringing economic doom. Don't blame the spotted owl for this.

3. The "Value" of Native Forests

Industrial forests provide wood, for a little while. This is their only value. *Native forests* provide wood, too, but they also provide homes for animals, plants, and insect life. They contain rich genetic and medicinal resources. They produce textile fibers and hundreds of other "products".

Forests preserve the soil against erosion. They slow the advance of deserts. They regulate climate: rainfall, humidity, temperature. Forests produce oxygen, mitigating global warming. They protect watersheds and fresh water resources. They are home to one of the planet's greatest sources of biological activity.

But there's still more to this story. Does it need to be repeated? *Native forests are glorious!* They are among the planet's most flamboyant, magnificent expressions of life.

And forests are teachers. They teach us the truths of nature's process, and our appropriate place within it. Forests are windows to the roots of creation. They inspire human imagination, and touch the human soul.

Forests are still more than this. They have an *intrinsic value* beyond objective measure. A society that sees them only as a resource to be exploited, as a crop to be marketed, has lost its sense of the sacred. Saving America's forests is more than an economic or an ecological issue. It is a spiritual one as well.

4. Some Principles of Ecoforestry

If our forests are to be saved for the long run, the reality of nature's limits must be respected. *This is the basis of ecoforestry.* Industrial forestry attempts to defy these limits, to push only one value—wood production—to its impossible maximum. We now know this can never work. It leaves the soil, air, water, weather, wildlife, genetic resources and ecosystems too damaged. In the end it destroys economies. That way is a path to failure.

There *is* a better way. Here are a few of its principles:

- We must accept that biological diversity—diverse species, genes, landscapes, communities—in all their natural patterns, is the basis of fully functioning, healthy ecosystems.

- All human activity must, first of all, be ecologically responsible. This standard must take precedent over perceived self-interest.

- Uses of a forest must be balanced, so that all organisms, human and non-human, are provided with a fair and protected land base; none are sacrificed for short-run human benefit.

- Clearcutting must be outlawed; it kills ecosystems like a bullet through a brain.

- Pesticide use must be banned; it poisons the land and humans.

- Ecological "planning" periods should not be less than 150–250 years.

- Wood may be harvested, but from the naturally occurring surplus product, thrown off by a fully functioning, ecologically balanced forest. Natural selection.

- Ecological succession must be maintained, to protect biodiversity; do not use "brush control."

- Road building and use must be minimal.

- Drainage systems, streams, lakes and rivers must be protected.

- Industrial logging jobs must be diverted to ecoforestry and restoration jobs, which are far more labor intensive, benefit the planet and community, and are constructive, not destructive.

5. Renewal is Labor Intensive

Renewal is a key element. If we could turn around the energy that is now devoted to destroying forests, communities suffering job losses in the current transition period in the Pacific Northwest would also be renewed.

Take the example of Redwood National Park. When the park was expanded in 1978, much of lower Redwood Creek valley had been brutalized by decades of industrial forestry. Ancient redwood groves along the creek were threatened by erosion. The Park Service developed one of the first large-scale plans for restoration. They hired many people to stop erosion, clean out small creekbeds, obliterate roads. Today the parklands are healing. Alders grow thick in areas once clearcut. Native grasses are returning to meadows where logging roads were "put to bed" and hillsides reshaped to their original slope. Elk, mountain lion, and black bear have returned. It may take centuries before huge redwoods grow in cutover areas, but nature has the time. If we allow it.

Protection of remaining ancient forests, restoration, renewal and *ecoforestry* will not only save forests from the ravages of the failed industrial model, they will also be *positive* models for humanity's relationship to nature. Forests *can* be sustained for the long run. But to reverse our present course will require acceptance of certain limits, and tough choices. For example, we must use fewer wood products and ban the export of raw logs. (If we export any wood products in the future, it should be after processing, with higher "values added," and greater labor inputs.) We must harvest only the *forest surplus*, determining this by standards of *natural selection forestry*, part of ecoforestry. Our universities must stop training plantation managers, and start training ecoforesters, who understand nature's ways. And we must eliminate tax breaks for corporations that cut forests, giving those breaks to ecologically responsible businesses.

We must adopt a new forestry that respects forests for their *intrinsic* worth, that places diversity and ecological health above other standards: one that looks less at the corporate bottom line and more at nature's bottom line. *This* is what we would like to hear from Mr. Clinton: *The end to industrial forestry models.*

There is much more to say, but insufficient room. If you would like more information about *ecoforestry* as a practical alternative to ruinous industrial forestry models, please write to: The Ecoforestry Institute of the US, PO Box 12543, Portland, OR 97212, or, in Canada to: The Ecoforestry Institute Society, PO Box 5783, Stn. B, Victoria, BC, Canada, V8R 6S8.